T0212416

Lecture Notes in Mathematics

Volume 2309

This series reports on new developments in all areas of mathematics and their applications - quickly, informally and at a high level. Mathematical texts analysing new developments in modelling and numerical simulation are welcome. The type of material considered for publication includes:

1. Research monographs
2. Lectures on a new field or presentations of a new angle in a classical field
3. Summer schools and intensive courses on topics of current research.

Texts which are out of print but still in demand may also be considered if they fall within these categories. The timeliness of a manuscript is sometimes more important than its form, which may be preliminary or tentative.

Titles from this series are indexed by Scopus, Web of Science, Mathematical Reviews, and zbMATH.

Alberto Torchinsky

A Modern View of the Riemann Integral

 Springer

Alberto Torchinsky
Department of Mathematics
Indiana University
Bloomington, IN, USA

ISSN 0075-8434   ISSN 1617-9692   (electronic)
Lecture Notes in Mathematics
ISBN 978-3-031-11798-5   ISBN 978-3-031-11799-2   (eBook)
https://doi.org/10.1007/978-3-031-11799-2

Mathematics Subject Classification: 26A42

This Springer imprint is published by the registered company Springer Nature Switzerland AG
The registered company address is: Gewerbestrasse 11, 6330 Cham, Switzerland

To
  Massi
  Lindsey, Cyrus, Zarina
  Shiqi, and Darius

Preface

The Riemann integral is the most familiar integral to mathematicians and those in fields that merely use mathematics alike. Now, the characterization of Riemann integrable functions relies on the notion of sets of Lebesgue measure zero, and a Riemann integrable function on a bounded interval is Lebesgue integrable there; thus, in a way, the Lebesgue integral is more general than the Riemann integral. On the other hand, the Lebesgue integral can be expressed as a Riemann–Stieltjes integral [80] or as an improper Riemann integral [41]. Moreover, there are instances when the Riemann integral is applicable whereas the Lebesgue integral is not, for example, when uniformly distributed sequences or improper integrals are involved.

These considerations give us pause for thought as to the role the Lebesgue theory plays in Riemann integration. R. W. Hamming has expressed his feelings in this matter forcefully, "...for more than 40 years I have claimed that if whether an airplane would fly or not depended on whether some function that arose in its design was Lebesgue *but not Riemann* integrable, then I would not fly in it. Would you? Does Nature recognize the difference? I doubt it!" [43].

As for us, in this book we introduce the tools necessary for developing the Riemann integral as seen through the lens of Riemann's original viewpoint, and show its handiness in computations and applications.

Our approach rests on three basic observations. The first is a Riemann integrability criterion in terms of oscillations, which is a quantitative formulation of the fact that Riemann integrable functions are continuous almost everywhere with respect to the Lebesgue measure. Second, we analyze the nature of the Riemann sums, and introduce the concepts of admissible families of partitions and modified Riemann sums; the latter is motivated by an experiment in signal retrieval [64] and allows for sets other than intervals in the definition of Riemann sums. Lastly, we exploit the fact that most numerical quadrature rules make use of carefully chosen Riemann sums, and thus the Riemann integral, be it proper or improper, is most appropriate for this endeavor.

The book is intended for mature graduate students, whom it will serve as an up-to-date primer, and researchers and teachers in mathematics and surrounds, whom it will serve as an advanced resource tome on Riemann integration and its applications. It draws only on a proof-based familiarity with the Riemann integral, though extends these concepts in several sophisticated directions. This unusual combination means it is potentially accessible to those enthusiasts interested in exploring the possibilities of Riemann's original notion of integral, including those intrepid undergraduates keen to glimpse a world without Lebesgue, though appropriate guidance would be required.

It is always a pleasure to acknowledge the contributions of those who make a project of this nature possible. The Tech Support Team was prompt and efficient. And, the book owes much to the vision and insight of Dr. Loretta Bartolini, who was the best editor this ambitious project could have had.

Bloomington, IN, USA   Alberto Torchinsky
February 2022

Contents

Chapter 1
Introduction

Bernhard Riemann submitted his *Habilitationsschrift*—the postdoctoral examination—the postdoctoral examination—the postdoctoral examination required to qualify to lecture at a university—to the University of Göttingen
at the end of 1853. He had spent 30 months working on the dissertation, and in
the fourth section, entitled "Ueber den Begriff eines bestimmten Integrals und den
Umfang seiner Gültigkeit" ("On a notion of a definite integral and the scope of its
validity"), Riemann introduced the following condition for a function to have an
integral on an interval.

Let $I = [a, b]$ be a bounded interval in $\mathbb{R}$, and let $(\mathcal{P}, C)$ be a *tagged partition*
of I, i.e., a partition $\mathcal{P} = \{I_1, \ldots, I_m\}$ of I together with a set $C = \{c_1, \ldots, c_m\}$
of points in I such that $c_k \in I_k$ for $1 \le k \le m$. Given a real-valued function f
defined on I, let $S(f, \mathcal{P}, C)$ denote the *Riemann sum* of f with respect to the tagged
partition $(\mathcal{P}, C)$ of I, i.e.,

$$S(f, \mathcal{P}, C) = \sum_{k=1}^{m} f(c_k) |I_k|. \tag{1}$$

Then, we say that f is *Riemann integrable* on I if there is a real number R with
the following property: For every $\varepsilon > 0$, there exists $\delta > 0$ such that $|S(f, \mathcal{P}, C) - R| \le \varepsilon$ for every tagged partition $(\mathcal{P}, C)$ of I with *mesh* $\|\mathcal{P}\| = \max_k\{|I_k|\} \le \delta$. In
that case, $\int_I f = R$ denotes the *Riemann integral* of f on I.

Earlier, in the twenty-first lecture of the influential *Résumé des leçons données a
l'École royale polytechnique sur le calcul infinitésimal* (Summary of Lectures given
at the Royal Polytechnic School on the Infinitesimal Calculus), published in 1823,
Augustin Cauchy had defined the definite integral of a continuous function f as the
limit of sums, rather than in terms of antiderivatives as had been traditional during
the seventeenth and eighteenth centuries [18].

Now, whereas Cauchy restricted the tags in the Riemann sums of f to the left
endpoints of the partition intervals, Riemann allowed for the free choice of the tags
within each interval [82]. D. C. Gillespie observed that restricting the tags in (1)

© The Author(s), under exclusive license to Springer Nature Switzerland AG 2022   1
A. Torchinsky, *A Modern View of the Riemann Integral*,
Lecture Notes in Mathematics 2309, https://doi.org/10.1007/978-3-031-11799-2_1

above to the left endpoints of the partition intervals defines the same integral [39], and generalizations of this observation have been given in [1, 55].

Due to its intuitive formulation as the "area under the curve" and its many applications outside of mathematics, the Riemann integral is perhaps the most familiar integral for mathematicians and non-mathematicians alike. And, although mathematicians may agree that Riemann's approach is an easy way to think about integration, they are quick to point out its limitations.

For one, there is the evaluation of the limit in (1). Carr and Hill observed that the existence of the limit and its value depend upon both the manner of subdivision of I and the restriction of the summation to prescribed subsets of $\{1, \ldots, m\}$, or *patterns*. They considered the case when I is subdivided into n equal subintervals, and the prescribed subsets form a *variable* or *fixed pattern* (defined below). In the latter case, the resulting limit, provided it exists, is the *pattern integral* of f, [17]. McShane noted that the sum is not a single-valued function of the number of subdivisions and questioned what kind of limit is taken [68]. Renz observed that the numerical treatment of (1) is awkward and that the limit of the Riemann sums, which involve arbitrary partitions and tag points, is complicated [79]. Chui examined the rate of convergence of the right and uneven Riemann sums of f to the integral of f when I is subdivided into n equal subintervals and noted the similarity with the behaviour of the sequence of the Fourier coefficients of f, [19, 20]. Bennett and Jameson considered conditions that ensure that the left, right, and other related Riemann sums of f, including the upper and lower sums, converge monotonically to the integral of f when I is subdivided into n equal subintervals [8]. Sklar restricted the partitions to a regular sequence and defined the Riemann integral sequentially [90]. He also introduced the *uniform integral* of f, which is defined as an ordinary numerical limit and which coincides with the Riemann integral of f when f is Riemann integrable, [85, 106]. Earlier, Bromwich and Hardy made use of this concept in dealing with the representation of *improper* Riemann integrals as a limit of infinite Riemann sums [12], and Wintner [108] and Ingham [48] used it in the proof of the prime number theorem. Osgood and Shisha introduced the *dominated* (and the *simple*) integral in order to apply quadrature formulas to the numerical evaluation of improper Riemann integrals. The dominated integral of f is defined by means of a single limit, and when it exists, it agrees with the improper Riemann integral of f, [72].

In this monograph, we will address these observations within the framework of the theory of Riemann integration in a simple, transparent manner, one that is adaptable to computations. To accomplish this, we will introduce the notion of *admissible family* Π of partitions of I and the associated Π-*Riemann integral* on I. We will prove that if the partitions in (1) are limited to an admissible family, the resulting Π-integral coincides with the Riemann integral on I. As for the summation, as in the Lebesgue integral, we will allow for sets other than intervals in the partitions. This we will achieve by means of the *modified Riemann sums* of f, [100].

A particular instance of these sums was first considered in an experiment in signal retrieval; we describe a simplified version of the physical situation in EE terms [64]. One of the most common means of information transmission is amplitude modulation (AM), where a signal of frequency f is multiplied by a carrier wave at a much higher frequency f_c, permitting signal transmission over long distances. Generally, the receiver recovers the originally modulated signal by multiplying the received waveform by a "local oscillator", i.e., a local copy of the carrier wave at f_c, and then filtering away the high-frequency components, forming the basis of AM radio. Occasionally, however, direct analysis of the signal itself is required rather than first demodulating the carrier wave, e.g., when a local oscillator field is not available. In these cases, it becomes necessary to develop an accurate treatment of the original signal itself.

When the signals are modulated by a carrier square wave of unit amplitude, as may occur in fast digitization of modulated signals, two considerations come into play. First, of course, Gibbs' phenomenon [33]. And, second, the realization that the operations on the original signal correspond to the integration of a continuous function f over a domain where the support of f is periodically reduced by a factor of 1/2. In terms of the approximating Riemann sums of the integral of f, this translates into halving each interval that appears in the Riemann sums $\sum_{k=1}^{m} f(x_k)|I_k|$ of f, and considering modified Riemann sums of the form $\sum_{k=1}^{m} f(x_k^1)|I_k^1|$, where each x_k^1 lies in the left half I_k^1 of I_k, $1 \leq k \leq m$. That these sums converge to $(1/2) \int_I f$ was derived experimentally in [64], and the result will be revisited in its general form in Examples 1, 4, 5, and 6 below.

Our approach relies entirely on basic properties of the Riemann integral. The subjects covered fall into three broad categories: quadrature rules, modified Riemann sums, and improper integrals. The results are presented in six chapters, each with a short introduction and applications and examples, ranging from Fourier series to uniformly distributed sequences, and from calculating the volume of a lake to the quadrature evaluation of improper integrals.

In Chap. 2, we introduce the Π-Riemann integral. A quantitative formulation of the principle that Riemann integrable functions on an interval are continuous a.e. with respect to the Lebesgue measure there, formulated in terms of oscillations in (6) below [13, 46], is an important tool in dealing with Π-integrable functions. We also show that even though the Riemann sums of an integrable function converge to the integral, they may do so arbitrarily slowly. We then discuss various numerical methods, or *quadrature rules*, to evaluate a definite integral. Since most numerical quadrature rules make use of carefully chosen Riemann sums, the Riemann integral is most appropriate for this endeavour. And we observe that if the function in question is convex, or concave, the upper and lower sums converge to the integral monotonically.

In Chap. 3, we prove a basic convergence theorem for the Riemann integral. Although the Riemann integral does not enjoy satisfactory properties with respect to pointwise limits, Theorem 13 allows for the taking of limits. In particular, it gives the Riemann–Lebesgue lemma of Fourier series in its various formulations. We also cover the Weierstrass algebraic and polynomial approximation theorems.

In Chap. 4, we introduce the modified Π-Riemann sums. Some of the applications discussed include ψ-asymptotically distributed sequences, uniformly distributed sequences, and extensions of recent results on deleting items and disturbing mesh in the Riemann integral [61].

In Chap. 5, we introduce the pattern and uniform integrals. Whereas the original applications of the pattern integrals were limited to summability methods [45], we are interested in sequences more general than patterns, the modified Π-Riemann sums being the natural setting for these results.

In Chap. 6, we discuss the improper and dominated integrals; in particular, the latter addresses the question as to when a function is majorized by a positive steadily decreasing function with a convergent improper integral. We also explore the relation between uniform and improper integrals and note the simplicity of computation when dealing with uniform integrals; this is true in other contexts as well. Quadrature rules converge to the integral of the function when the function is Riemann integrable, but in the case of the improper Riemann integral, this connection is obscured by the double limiting process involved. The dominated integral, which is defined by a single limit, allows for this connection.

In Chap. 7, the Coda, we discuss briefly the extension of our results to the context of Riemann–Stieltjes integrals [100]. Applications of these results to stochastic integrals, for instance, are discussed in [60], and we also touch briefly on modified Riemann sums, asymptotic distribution functions, and quadrature formulas. The extension of our results to this setting is natural.

Finally, in Appendix I, we cover the change of variable formulas for Riemann and Riemann–Stieltjes integrals, closing with the caveat that not always the most general results are the most useful. And, in Appendix II, we come full circle and prove that bounded functions that are integrable à la Cauchy are Riemann integrable and that the integrals coincide.

The notation is standard, or introduced as we go along. We only point out that for sequences $\alpha_n > 0$ and A_n, $A_n = O(\alpha_n)$ means that $|A_n|/\alpha_n$ is bounded by a constant that may depend on the various quantities appearing in the expression but is independent of n.

Chapter 2
The Π-Riemann Integral

In this chapter, we will introduce the Π-Riemann integral and discuss its basic properties. In particular, we observe that even though the Riemann sums of an integrable function converge to the integral, they may do so arbitrarily slowly. It is often the case, e.g., when computing the volume of a lake, that we take the limit of sums closely related to the Riemann sums of f, and we are thus led to Bliss' theorem. We also discuss various numerical methods, or quadrature rules, to evaluate definite integrals, with special emphasis on the rate of approximation. The methods include the right and left Riemann sums, the uneven tag sums, and the midpoint, trapezoid, and Simpson rules. These methods also allow for the computation of roots of nonlinear equations. And we observe that if the function in question is convex, or concave, the upper and lower sums converge to the integral monotonically.

We begin by introducing the families of partitions that constitute the basis for our discussion. Let $I = [a, b]$ be a bounded interval in $\mathbb{R}$. A family $\Pi = \{\mathcal{P}\}$ of partitions of I is said to be *admissible* provided it satisfies the following two properties:

(i) Given $\eta > 0$, there is a partition $\mathcal{P}$ of I in Π with mesh $\|\mathcal{P}\| \leq \eta$.
(ii) If $\mathcal{P}, \mathcal{P}_1$ are partitions of I in Π, there is a partition $\mathcal{P}_2$ of I in Π that is a common refinement of $\mathcal{P}$ and $\mathcal{P}_1$.

Instances of admissible families include the collection of all partitions of I; Π_e, the partitions of I that result by dividing I into n equal length subintervals, all n; the partitions of I that result by dividing I into $2n + 1$ equal length subintervals, all n; the (nested) dyadic partitions of I; and partitions that result by dividing I into k^n equal length subintervals, all n, where k is an integer > 1.

Given a real-valued function f defined on I, we introduce the Π-*Riemann integral* of f as follows. Let $S_\Pi(f, \mathcal{P}, C)$ denote the Riemann sum of f with respect to the tagged partition $(\mathcal{P}, C)$ of I given in (1) where $\mathcal{P}$ is restricted to Π. We say that f is Π-*Riemann integrable on* I if there is a real number R such that, for every

© The Author(s), under exclusive license to Springer Nature Switzerland AG 2022
A. Torchinsky, *A Modern View of the Riemann Integral*,
Lecture Notes in Mathematics 2309, https://doi.org/10.1007/978-3-031-11799-2_2

$\varepsilon > 0$, there exists $\delta > 0$ such that $|S_\Pi(f, \mathcal{P}, C) - R| \le \varepsilon$ for every tagged partition $(\mathcal{P}, C)$ of I, $\mathcal{P} \in \Pi$, with $\|\mathcal{P}\| \le \delta$. In that case, R denotes the Π-Riemann integral of f on I.

Along similar lines, we introduce the Π-Darboux integral of f. Given a partition $\mathcal{P}$ of I, $\mathcal{P} = \{I_1, \ldots, I_m\}$, let

$$M_k = \sup_{I_k} f, \quad \text{and,} \quad m_k = \inf_{I_k} f, \quad 1 \le k \le m.$$

For a partition $\mathcal{P} = \{I_1, \ldots, I_m\}$ of I in Π and a bounded function f on I, let $U_\Pi(f, \mathcal{P})$ and $L_\Pi(f, \mathcal{P})$ denote the upper and lower Riemann sums of f along $\mathcal{P}$ on I, i.e.,

$$U_\Pi(f, \mathcal{P}) = \sum_{k=1}^m \left(\sup_{I_k} f\right) |I_k|, \quad \text{and,} \quad L_\Pi(f, \mathcal{P}) = \sum_{k=1}^m \left(\inf_{I_k} f\right) |I_k|,$$

respectively, and set

$$U_\Pi(f) = \inf_{\mathcal{P} \in \Pi} U_\Pi(f, \mathcal{P}), \quad \text{and,} \quad L_\Pi(f) = \sup_{\mathcal{P} \in \Pi} L_\Pi(f, \mathcal{P}).$$

We then say that f is Π-*Darboux integrable* on I if $U_\Pi(f) = L_\Pi(f)$, and in this case, the common value is the Π-Darboux integral of f on I.

We first observe that both Π-integrals coincide. In fact, it turns out that f is Riemann integrable, and the common value of the Π-integrals is precisely $\int_I f$.

Theorem 1 *Let I be a bounded interval of $\mathbb{R}$, Π an admissible family of partitions of I, and f a function defined on I. Then, f is Π-Riemann integrable on I iff f is Π-Darboux integrable on I. Furthermore, f is Riemann integrable on I, and the value of both integrals coincides with $\int_I f$, the Riemann integral of f on I.*

Proof Assume first that f is Π-Riemann integrable on I with integral R, and note that f is bounded on I. Indeed, pick the $\delta > 0$ corresponding to $\varepsilon = 1$ in (1), fix a partition $\mathcal{P} = \{I_1, \ldots, I_m\}$ of I in Π with mesh $\|\mathcal{P}\| \le \delta$ (which exists by (i)), and consider a tag set $C = \{c_1, \ldots, c_m\}$ in I with $c_k \in I_k$, $1 \le k \le m$. Keeping the c_k in C fixed except for c_{k_0}, and letting it vary in I_{k_0}, it follows from (1) that f is bounded on I_{k_0} and, hence, on all of I.

Let now $\varepsilon > 0$, pick a partition $\mathcal{P} = \{I_1, \ldots, I_m\}$ of I in Π such that $|S_\Pi(f, \mathcal{P}, C) - R| \le \varepsilon/2$ for every set of tags $C = \{c_k \in I_k : 1 \le k \le m\}$, and pick $c_k \in I_k$ such that $M_k \le f(c_k) + \varepsilon/2|I|$ for $1 \le k \le m$. It then follows that $U_\Pi(f, \mathcal{P}) = \sum_k M_k |I_k| \le \sum_k f(c_k)|I_k| + \varepsilon/2 = S_\Pi(f, \mathcal{P}, C) + \varepsilon/2$, which, since $S_\Pi(f, \mathcal{P}, C) \le R + \varepsilon/2$, gives

$$U_\Pi(f) \le U_\Pi(f, \mathcal{P}) \le R + \varepsilon. \tag{2}$$

Moreover, since $-L_\Pi(f, \mathcal{P}) = U_\Pi(-f, \mathcal{P})$ and the value R that corresponds to f is equal to the value $-R$ corresponding to $-f$, (2) applied to $-f$ gives $-L_\Pi(f, \mathcal{P}) = U_\Pi(-f, \mathcal{P}) \le -R + \varepsilon$ or

$$R \le L_\Pi(f, \mathcal{P}) + \varepsilon \le L_\Pi(f) + \varepsilon. \tag{3}$$

Thus, combining (2) and (3), we have that $U_\Pi(f) - \varepsilon \le R \le L_\Pi(f) + \varepsilon$, for every $\varepsilon > 0$, and, therefore, $U_\Pi(f) \le R \le L_\Pi(f)$. Since also $L_\Pi(f) \le U_\Pi(f)$, we conclude that $L_\Pi(f) = R = U_\Pi(f)$, and f is Π-Darboux integrable with integral R.

Conversely, suppose that f is Π-Darboux integrable with integral R. We claim that, given $\varepsilon > 0$, there is a partition Q of I in Π such that

$$U_\Pi(f, Q) - L_\Pi(f, Q) \le \varepsilon. \tag{4}$$

Indeed, since

$$\inf_{\mathcal{P} \in \Pi} U_\Pi(f, \mathcal{P}) = \sup_{\mathcal{P} \in \Pi} L_\Pi(f, \mathcal{P}) = R,$$

there are partitions $\mathcal{P}, \mathcal{P}_1$ of I in Π such that $U_\Pi(f, \mathcal{P}) \le R + \varepsilon/2$, and, $R \le L_\Pi(f, \mathcal{P}_1) + \varepsilon/2$. Now, by (ii), there is a partition Q of I in Π that is a common refinement of $\mathcal{P}$ and $\mathcal{P}_1$. Whence, since $L_\Pi(f, Q) \ge L_\Pi(f, \mathcal{P}_1)$ and $U_\Pi(f, Q) \le U_\Pi(f, \mathcal{P})$, it follows that

$$U_\Pi(f, Q) - L_\Pi(f, Q) = U_\Pi(f, Q) - R - \left(L_\Pi(f, Q) - R\right)$$
$$\le U_\Pi(f, \mathcal{P}) - R - \left(L_\Pi(f, \mathcal{P}_1) - R\right) \le \varepsilon,$$

and (4) holds.

Recall that by a familiar fact in Riemann integration, provided that (4) holds, f is Riemann integrable on I and, in particular, (1) holds [13, 46]. When restricted to Π, (1) gives that f is Π-Riemann integrable on I and that the Π-Riemann integral of f on I equals $\int_I f$. Since by the first part of the proof the Π-Riemann and the Π-Darboux integrals of f on I coincide, the proof is finished. □

We will require another basic result, namely, a characterization of Π-integrability in terms of the oscillation of a function [13, 46]. Recall that given a bounded function f defined on I and an interval $J \subset I$, the *oscillation* osc(f, J) *of* f *on* J is defined as osc$(f, J) = \sup_J f - \inf_J f$.

We then have:

Proposition 1 *Let I be a bounded interval in $\mathbb{R}$, Π an admissible family of partitions of I, and f a Π-Riemann integrable function defined on I. Then, there is a family $\{\mathcal{P}_n\}$ of partitions of I in Π, $\mathcal{P}_n = \{I_1^n, \ldots, I_{m_n}^n\}$, such that*

$$\lim_n \left(U_\Pi(f, \mathcal{P}_n) - L_\Pi(f, \mathcal{P}_n)\right) = 0, \tag{5}$$

$$\lim_n \sum_{k=1}^{m_n} \operatorname{osc}(f, I_k^n) \, |I_k^n| = 0, \tag{6}$$

and

$$\lim_n U_\Pi(f, \mathcal{P}_n) = \lim_n L_\Pi(f, \mathcal{P}_n) = \int_I f. \tag{7}$$

Conversely, if (6) holds, f is Π-Riemann integrable on I.

Proof First, since f is Π-Riemann integrable on I with integral R, given $\varepsilon > 0$, there exists $\delta > 0$ such that $|S_\Pi(f, \mathcal{P}, C) - R| \le \varepsilon/4$ for all tagged partitions $(\mathcal{P}, C)$ of I in Π with mesh $\|\mathcal{P}\| \le \delta$. In particular, $|S_\Pi(f, \mathcal{P}, C_1) - S_\Pi(f, \mathcal{P}, C_2)| \le \varepsilon/2$ for any partition $\mathcal{P}$ of I in Π with mesh $\le \delta$ and tags C_1, C_2. Pick next the tags C_1, C_2 so that $U_\Pi(f, \mathcal{P}) \le S_\Pi(f, \mathcal{P}, C_1) + \varepsilon/4$ and $S_\Pi(f, \mathcal{P}, C_2) \le L_\Pi(f, \mathcal{P}) + \varepsilon/4$. Then it follows that

$$U_\Pi(f, \mathcal{P}) - L_\Pi(f, \mathcal{P})$$
$$\le U_\Pi(f, \mathcal{P}) - S_\Pi(f, \mathcal{P}, C_1) + S_\Pi(f, \mathcal{P}, C_2) - L_\Pi(f, \mathcal{P})$$
$$+ \left| S_\Pi(f, \mathcal{P}, C_1) - S_\Pi(f, \mathcal{P}, C_2) \right|$$
$$\le \varepsilon/4 + \varepsilon/4 + \varepsilon/2 = \varepsilon. \tag{8}$$

Fix N such that $1/N \le \delta$, and let $n \ge N$. By (i), there is a sequence $\{\mathcal{P}_n\}$ of partitions of I in Π with mesh $\|\mathcal{P}_n\| \le 1/n \le \delta$, and by (8), it follows that

$$U_\Pi(f, \mathcal{P}_n) - L_\Pi(f, \mathcal{P}_n) \le \varepsilon, \quad \text{all } n \ge N,$$

and (5) holds. Moreover, since

$$U_\Pi(f) - L_\Pi(f) \le U_\Pi(f, \mathcal{P}_n) - L_\Pi(f, \mathcal{P}_n) \le 1/n, \quad \text{all } n,$$

$U_\Pi(f) = L_\Pi(f)$ and (7) holds. Finally, since

$$\sum_{k=1}^{m_n} \operatorname{osc}(f, I_k^n) \, |I_k^n| = U_\Pi(f, \mathcal{P}_n) - L_\Pi(f, \mathcal{P}_n) \le \varepsilon, \quad \text{all } n \ge N,$$

the above sum is $\le \varepsilon$ for all $n \ge N$, (6) holds. The proof of the first statement is thus complete.

Conversely, let $\varepsilon > 0$ be given. Then from (6) and the relation above, it follows that for sufficiently large n there is a partition $\mathcal{P}_n$ of I in Π such that $U_\Pi(f, \mathcal{P}_n) - L_\Pi(f, \mathcal{P}_n) \le \varepsilon$, and, consequently, (4) holds, and by Theorem 1, f is Riemann integrable on I. $\qquad \square$

Condition (6) is a quantitative formulation of the fact that Riemann integrable functions on an interval are continuous a.e.with respect to the Lebesgue measure there. In his dissertation, Riemann also introduced the following integrability criterion in terms of oscillations. Given a bounded function f on an interval I, a fixed $\eta > 0$, and a partition $\mathcal{P} = \{J_1, \ldots, J_m\}$ of I, let $A_\eta(\mathcal{P}) = \{k : 1 \le k \le m$, and osc $(f, J_k) \ge \eta\}$. Then, with

$$s_\eta(\mathcal{P}) = \sum_{k \in A_\eta(\mathcal{P})} |I_k|,$$

f is integrable on I iff for each fixed $\eta > 0$, $s_\eta(\mathcal{P}) \to 0$ as $\|\mathcal{P}\| \to 0$. That is, for every fixed $\eta > 0$, given $\varepsilon > 0$, there is $\delta > 0$ such that $s_\eta(\mathcal{P}) \le \varepsilon$ whenever $\|\mathcal{P}\| \le \delta$.

Based on this criterion, Riemann concluded that the function f given by the convergent series

$$f(x) = \frac{(x)}{1^2} + \frac{(2x)}{2^2} + \cdots + \frac{(nx)}{n^2} + \cdots,$$

where (x) denotes the (positive or negative) distance to the nearest integer function, is integrable. Now, f is discontinuous on the dense set of points $x = m/2n$, where m, n are relatively prime. At these points, the left- and right-hand side limiting values of f are

$$f\left(\left(\frac{m}{2n}\right)^-\right) = f\left(\frac{m}{2n}\right) + \frac{1}{2n^2}\left(1 + \frac{1}{3^2} + \frac{1}{5^2} + \cdots\right) = f\left(\frac{m}{2n}\right) + \frac{\pi^2}{16\,n^2}$$

and

$$f\left(\left(\frac{m}{2n}\right)^+\right) = f\left(\frac{m}{2n}\right) - \frac{1}{2n^2}\left(1 + \frac{1}{3^2} + \frac{1}{5^2} + \cdots\right) = f\left(\frac{m}{2n}\right) - \frac{\pi^2}{16\,n^2},$$

respectively, and so the number of discontinuities in any given interval is infinite, but the number of discontinuities, which in any given interval exceed a given $\eta > 0$, is always finite.

Riemann discussed the notion of integral in less than six pages of his dissertation and restricted the analysis to its characterization. Riemann's thesis, a handwritten manuscript preserved at the Göttingen University Library Archives, was first published posthumously in 1868 due to the efforts of R. Dedekind, who arranged for it to appear in the Abhandlungen der Königlichen Gesellschaft der Wissenschaften zu Göttingen (Treatises of the Göttingen Royal Society of Sciences); it is likely that until then only a handful of mathematicians had access to its content or perhaps had heard about it. H. Hankel showed in 1870 that Riemann integrable functions are necessarily continuous on a dense set of points in their domain of integration [44]. And in 1875, H. J. S. Smith pointed out that further discussion of Riemann's results would seem to be desirable partly because some demonstrations were wanting in formal accuracy and partly because the theorems themselves appeared to have been misunderstood and to have been made the basis of erroneous inferences. It was precisely to clarify some of Hankel's results that Smith introduced a kind of sets that allow for a function to be discontinuous, yet integrable; these sets essentially include the Cantor sets, both of measure zero and of positive measure. In Smith's definition, the "last segment" of an interval is removed, while in the Cantor sets, the "middle" third segment is removed [92].

We will first consider Hankel's result. Given a bounded function f defined on a closed bounded interval I and an interior point x of I, by the *oscillation* osc (f, x) *of f at x*, we mean

$$\text{osc}\,(f, x) = \lim_{\eta \to 0^+} \text{osc}\,(f, [x - \eta, x + \eta]),$$

with the usual adjustment when x is an endpoint of I.

It is then readily verified that for interior points x of I, f is continuous at x iff osc $(f, x) = 0$. Indeed, suppose first that f is continuous at x and let $\varepsilon > 0$. Then there exists $\delta > 0$ such that if $y \in (x - \delta, x + \delta) \subset I$, then $|f(x) - f(y)| \leq \varepsilon/2$, and, consequently, $|f(y) - f(z)| \leq \varepsilon$ whenever $y, z \in (x - \delta, x + \delta)$. Therefore, osc $(f, [x - \delta, x + \delta]) \leq \varepsilon$, and since ε is arbitrary, osc $(f, x) = 0$.

Conversely, suppose that osc $(f, x) = 0$, and let $\varepsilon > 0$. Then there exists $\delta = \delta(\varepsilon) > 0$ such that if $0 < h < \delta$, then $(x - h, x + h) \subset I$, and

$$\sup\{|f(y) - f(z)| : y, z \in (x - h, x + h)\} \leq \varepsilon.$$

Thus, if $|x - y| < \delta$ and $x, y \in I$, then $|f(x) - f(y)| \leq \varepsilon$, and f is continuous at x.

The proof of Hankel's result presented here rests on Cantor's nested interval theorem that asserts that the intersection of a nested sequence of nonempty closed subintervals of a closed bounded interval I is nonempty. We then have:

Proposition 2 *Let f be Darboux integrable on a bounded interval I. Then, $C_f = \{x \in I : f$ is continuous at $x\}$ is dense in I.*

Proof We will prove that if O is an open interval such that $O \cap I \neq \emptyset$, then $O \cap C_f \neq \emptyset$. So, given one such open interval O, let J be a nonempty closed interval such that $J \subset O \cap I$; then f is Darboux integrable on J. Let now $\{\varepsilon_n\}$ be a decreasing sequence that tends to 0. On account of (6), there is a partition $\mathcal{P}_1 = \{J_k^1\}$, $1 \leq k \leq n_1$, of J such that

$$U(f, \mathcal{P}_1) - L(f, \mathcal{P}_1) = \sum_{k=1}^{n_1} \text{osc}\,(f, J_k^1)\,|J_k^1| \leq |J|\,\varepsilon_1.$$

We claim that the oscillation of f on one of the intervals of $\mathcal{P}_1$ is $\leq \varepsilon_1$. Indeed, were this not the case we would have osc $(f, J_k^1) > \varepsilon_1$ for all the intervals of $\mathcal{P}_1$, and so, summing it would follow that $\sum_{k=1}^{n_1} \text{osc}\,(f, J_k^1)\,|J_k^1| > |J|\,\varepsilon_1$, which is not the case. We separate an interval where the oscillation does not exceed ε_1, rename it J_1, and note that $J \supseteq J_1$, and osc $(f, J_1) \leq \varepsilon_1$.

So, having chosen a nested collection of nonempty closed subintervals $J_1 \supseteq \ldots \supseteq J_n$ of J, with osc $(f, J_k) \leq \varepsilon_k$, $1 \leq k \leq n$, repeating the above argument with J_n in place of J and ε_{n+1} in place of ε_1, we find a nonempty closed interval $J_{n+1} \subseteq J_n$ such that osc $(f, J_{n+1}) \leq \varepsilon_{n+1}$, and so on. By Cantor's nested theorem, there is $x \in \bigcap_{n=1}^{\infty} J_n \subseteq J \subset I \cap O$; we claim that f is continuous at x. Indeed, since osc $(f, x) \leq$ osc $(f, J_k) \leq \varepsilon_k \to 0$ as $k \to \infty$, it follows that osc $(f, x) = 0$, f is continuous at x, and $O \cap C_f \neq \emptyset$.

Moreover, since dense sets are infinite, C_f is an infinite set and f is continuous at infinitely many points of I. $\qquad\square$

As for the Cantor set, or Cantor discontinuum, it is defined as follows. From $C_0 = I = [0, 1]$, we remove the open middle third interval and are left with $C_1 = [0, 1/3] \cup [2/3, 1]$, and upon removing the open middle third of each of the intervals in C_1, we are left with $C_2 = [0, 1/9] \cup [2/9, 1/3] \cup [2/3, 7/9] \cup [8/9, 1]$, and so on. Thus, in general, C_n is the union of 2^n closed intervals each of length $1/3^n$, and we let $C = \bigcap_{n=0}^{\infty} C_n$. Then C is closed and uncountable [99].

We claim that $f = \chi_C$ is Riemann integrable. To see this, let $f_n = \chi_{C_n}$ and observe that $f(x) \leq f_n(x)$ throughout I and that f_n is Riemann integrable on I with $\int_I f_n = (2/3)^n$. Now, given $\varepsilon > 0$, pick n so that $(2/3)^n \leq \varepsilon/2$, and choose a partition $\mathcal{P}$ of I such that $0 \leq U(f_n, \mathcal{P}) - \int_I f_n \leq \varepsilon/2$. In particular, $U(f_n, \mathcal{P}) \leq \varepsilon/2 + \int_I f_n \leq \varepsilon$, and since $f(x) \leq f_n(x)$ throughout I, also $U(f, \mathcal{P}) \leq U(f_n, \mathcal{P}) \leq \varepsilon$. Finally, since $L(f, \mathcal{P}) \geq 0$, it follows that $U(f, \mathcal{P}) - L(f, \mathcal{P}) \leq \varepsilon$ and, by (4), f is Riemann integrable. Moreover, since $\varepsilon > 0$ is arbitrary and $U(f, \mathcal{P}) \leq \varepsilon$, $\int_I f = 0$.

2.1 The Volume of a Lake

It is often the case that setting up and interpreting a problem require the taking of limits of sums closely related to the Riemann sums of a function f. Duhamel dealt with this situation by means of an interplay of infinitesimals; the background and applications of his approach are discussed by Osgood [74] and Ettlinger [32]. Bliss cast Duhamel's manipulations in mathematical terms and made use of the uniform continuity of continuous functions on a closed bounded interval to address it [9]. In our setting, we have:

Bliss' Theorem *Let $f_1, f_2, \ldots, f_N$ be a finite collection of Riemann integrable functions defined on a closed bounded interval $I = [a, b]$. Then $\prod_{i=1}^{N} f_i$ is Riemann integrable on I, and for any sequence $\mathcal{P}_m = \{J_k^m\}$, $1 \leq k \leq m_n$, $m \geq 1$, of partitions of I such that $\lim_m \|\mathcal{P}_m\| = 0$, and collections $\{c_{k,j}^m\}$ of tags with $1 \leq j \leq N$ and $1 \leq k \leq m_n$, such that $c_{k,j}^m \in J_k^m$, we have*

$$\lim_m \sum_{k=1}^{m_n} f_1(c_{k,1}^m) f_2(c_{k,2}^m) \cdots f_N(c_{k,N}^m) |J_k^m| = \int_I \prod_{i=1}^{N} f_i.$$

Proof The proof is by induction on the number of functions N. The case $N = 1$ is obvious. Now, clearly for $N > 1$, the product of the N functions is Riemann integrable. So suppose the statement is valid for $N - 1$ functions, and note that for m and the corresponding $1 \leq k \leq m_n$, we have

$$\prod_{i=1}^{N} f_i(c_{k,i}^m) = f_1(c_{k,1}^m) \Big(\prod_{i=2}^{N-1} f_i(c_{k,i}^m) \Big) f_N(c_{k,N}^m)$$

$$= \Big(\prod_{i=2}^{N-1} f_i(c_{k,i}^m) \Big) f_1(c_{k,1}^m) f_N(c_{k,1}^m)$$

$$+ \Big(\prod_{i=2}^{N-1} f_i(c_{k,i}^m) \Big) f_1(c_{k,1}^m) \Big(f_N(c_{k,N}^m) - f_N(c_{k,1}^m) \Big). \tag{9}$$

Since all functions are Riemann integrable and hence bounded, and since $c_{k,1}^m, c_{k,N}^m \in J_k^m$, with M the maximum among the products of the bounds for any $N - 1$ functions, the second summand in (9) does not exceed M osc (f_N, J_k^m), and, consequently,

$$\limsup_m \Big| \sum_{k=1}^{m_n} \Big(\prod_{i=2}^{N-1} f_i(c_{k,i}^m) \Big) f_1(c_{k,1}^m) \Big(f_N(c_{k,N}^m) - f_N(c_{k,1}^m) \Big) |J_k^m| \Big|$$

$$\leq M \limsup_m \sum_{k=1}^{m_n} \text{osc} \, (f_N, J_k^m) |J_k^m| = 0.$$

Hence, by the inductive assumption from (9), it readily follows that

$$\lim_m \sum_{k=1}^{m_n} \prod_{i=1}^{N} f_i(c_{k,i}^m) |J_k^m| = \lim_m \sum_{k=1}^{m_n} \Big(\prod_{i=2}^{N-1} f_i(c_{k,i}^m) \Big) f_1(c_{k,1}^m) f_N(c_{k,1}^m) |J_k^m|$$

$$= \int_I \Big(\prod_{i=2}^{N-1} f_i \Big) f_1 \, f_N = \int_I \prod_{i=1}^{N} f_i,$$

and the proof is finished.   □

Bliss' theorem applies to the evaluation of volume integrals, like the ones used in hydrocarbon reservoirs [66], earthwork computations [26], and the computation of the volume of a lake [35]. In such cases, structures are often irregular, and the volumes cannot be found by elementary integration techniques. Hence, the use of approximate integration, including a mechanical integrator, or planimeter, and approximation formulas when necessary.

One such formula is the *prismoidal formula*, [11, 70], which allows for the calculation of the volume of solids such as prismoids and prismatoids. Recall that a *prismatoid* is a polyhedron (a solid bounded by planes) with vertices that all lie in one or the other of two parallel planes. The two faces of the prismatoid lying in these planes are its bases; they do not necessarily have the same number of sides. A *prismoid* is a prismatoid with both bases having the same number of sides, and the lateral faces are quadrilaterals (either trapeziums or parallelograms). Special cases of a prismoid are a prism, with identical bases, and the frustum of a pyramid, i.e., a portion of a right regular pyramid included between the base and a section parallel to the base.

Now, the volume of a regular frustum of height h is given by the expression

$$V = \frac{1}{3}\left(A_1 + A_2 + \sqrt{A_1 A_2}\right)h,$$

where A_1 denotes the area of the lower base and A_2 that of the upper base. This is a particular instance of the prismoidal formula, deduced by Isaac Newton among others (his proof is given in [26]), which applies to any prismatoid solid, and is given by

$$V = \frac{1}{6}(A_1 + 4A_2 + A_3)\, h,$$

where A_1, A_3 are the areas of the bases at a distance apart of h, and A_2 is the area of the cross-section parallel and halfway between the end planes. This formula does in fact give the exact volume of a prismatoid, prismoid, prism, pyramid, frustum of a pyramid, wedge, cylinder, cone, frustum of a cone, sphere, ellipsoid, spherical segment, and, in general, any solid bounded by a quadric surface and two parallel planes.

These formulas have an interesting history. One of the oldest mathematical documents in existence, dating to around 1890 BC, is the Moscow Papyrus, an Egyptian papyrus roll approximately 5.44 m in length and 8 cm in height, consisting of one big piece with thirty-eight columns of text, and nine little fragments. The Moscow papyrus contains twenty-five problems with solutions, the first three being badly damaged, and the fourteenth posing the most intriguing problem of the calculation of the volume of the frustum of a truncated pyramid. The text, which extends over three columns (XXVII–XXIX) of the papyrus, begins by stating the problem, and the truncated pyramid is denoted by a drawing within the text. This is followed by the numerical data, the lengths of the lower side (4), upper side (2), and height (6). The procedure follows, one instruction is given, its result announced, and the next instruction follows until the final result (56) is arrived at [47].

The method of finding the volume as given in the papyrus can be expressed by the formula

$$V = \frac{h}{3}(a^2 + ab + b^2),$$

with a, b, and h denoting the measure of the sidelengths of the upper side, lower side, and height, respectively. The formula can be rewritten as

$$V = \frac{h}{6}\left(a^2 + 4\left(\frac{a+b}{2}\right)^2 + b^2\right),$$

which corresponds to the prismoidal formula. This brings to mind the familiar expression

$$\int_{[a,b]} f \approx \frac{(b-a)}{6}\left(f(a) + 4f\left(\frac{a+b}{2}\right) + f(b)\right), \tag{10}$$

which holds with an equality sign for polynomials f of degree less than or equal to three, and which expresses the approximation by means of the (simple) Simpson formula, which we will discuss below.

Various explanations have been advanced as to how the formula to calculate the volume of a truncated pyramid could have been discovered by the ancient Egyptians. Some use the knowledge of the calculation of volumes of other bodies, such as cylinders and cubes, and others a clever manipulation of algebraic formulas [40]. The level of sophistication required in arriving at the formula is higher than that for the rest of the problems posed in the papyrus, or that is otherwise used in the Egyptian mathematics of the time, leading some mathematicians to venture that the ancient Egyptians just stumbled into this particular formula, or perhaps it was part of a more advanced body of knowledge not reflected in the papyrus.

According to tradition, the god Thoth, originally associated with time and the Moon, gave the calendar, astronomy, and mathematics to the Egyptians. Thoth was later identified with the Greek god Hermes, who became Hermes Trismegistus (Thrice Great One), the presumed author of the Corpus Hermeticum, a set of writings alleged to have been transmitted from ancient times, where the secrets and techniques of influencing the stars and forces of nature were revealed.

Many of the seventeenth century scientists looked to magic and its allied arts, such as astrology, in celebrating the ideas and practical aspect of the Hermetic writings. Newton placed great faith in the idea of a "pure tradition", an unadulterated ancient doctrine that he studied earnestly, attempting to aid in his understanding of the physical world. Newton's manuscripts—most still unpublished—detail his thorough study of the Corpus Hermeticum. Newton was a devotee of hermetic studies, and it seems that he believed that his own discoveries, such as calculus, universal gravitation, and others, had been in the body of secret knowledge handed down from Thoth.

As for the three instances mentioned above, they all make use of the prismoidal formula. In the reservoir volume calculation, the trap anticline is approximated by frustum right circular cones, and the error is decreased by lowering the equidistance between integrated areas [66]. In the earthwork volume computations, as required for road construction, railroad embankments, and cuttings, dam construction, among others, the design is set out in the field, cross-section information obtained at regular intervals perpendicular to a centre line, and volumes computed from the cross-section areas and the interval distances [26].

And, to calculate the volume of a lake, limnologists—that is, scientists who study lakes—make use of a numerical integration method based on the volume of a cone frustum. A detailed map of the lake showing depth contours, i.e., the depth of the lake is the same at two points on a given contour, is drawn, and the function $A(x)$ that denotes the area contained in the contour at depth x is computed by means of a planimeter. A is defined on the interval $[0, d]$, where d is the maximum depth of the lake, and is strictly decreasing, and the volume V of the lake is given by $V = \int_{[0,d]} A$.

Now, A is not known exactly, and the integral must be approximated numerically. Since the slice of a lake between nearby horizontal planes is like a cone frustum, limnologists make use of the prismoidal approximation,

$$V \approx \sum_{k=1}^{n} \frac{A_{k-1} + A_k + \sqrt{A_{k-1} A_k}}{3} \Delta n,$$

where $A_k = A(x_k)$ for $k = 0, 1, \ldots, n$, are the areas of equally spaced depth contours and $\Delta n = d/n$ is the depth difference between consecutive contours.

More generally, given a continuous, nonnegative function f on an interval $[a, b]$, and a positive integer n, we define the *cone sum* $C_n(f)$ of f as

$$C_n(f) = \frac{(b-a)}{n} \sum_{k=1}^{n} \frac{f(x_{k-1}) + f(x_k) + \sqrt{f(x_{k-1})f(x_k)}}{3},$$

where $x_k = a + (b-a)k/n$ for $k = 0, \ldots, n$. If g is continuous, $\sqrt{g}$ is also continuous, and hence Riemann integrable on $[a, b]$. Then, by Bliss' theorem, the sum involving the third summand above converges to $(\int_I f)/3$, and so, by Bliss' theorem again, $\lim_n C_n(f) = \int_I f$. We may also reach this conclusion by observing that $C_n(f)$ lies between the lower and upper Riemann sums of f for appropriate partitions of I.

It is a practical concern to investigate the degree, or *rate of the approximation* of the cone sums $C_n(f)$ to the integral of f as $n \to \infty$. We are thus interested in the behaviour of

$$E_n(f) = \int_{[a,b]} f - C_n(f), \quad \text{as } n \to \infty.$$

We then have:

Proposition 3 *Let f be a nonnegative twice differentiable function defined on a closed bounded interval $I = [a, b]$. Then, with M_0 and M_2 the bounds for $f^{1/2}$ and*

$f^{1/2}{}''$, respectively, we have

$$\left| E_n(f) \right| \leq \frac{1}{6} M_0 M_2 \frac{(b-a)^3}{n^2}, \quad n = 1, 2, \ldots$$

Proof We will first consider the case $n = 1$. Since $f \geq 0$, we let $f = g^2$ and put

$$E_1(g, x) = \int_{[a,x]} g^2 - (x-a) \frac{g^2(x) + g^2(a) + g(x)g(a)}{3}, \quad a < x \leq b.$$

A straightforward computation then gives

$$E_1'(g, x) = \frac{1}{3}\big(2g(x) + g(a)\big)\big(g(x) - g(a) - (x-a)\,g'(x)\big), \quad a < x \leq b.$$

Now, if a function g is twice differentiable on $[a, b]$, by a simple application of the mean value theorem for derivatives, it follows that for $a < x \leq b$, we have

$$g(x) = g(a) + (x-a)g'(a) + \frac{1}{2}(x-a)^2 g''(c),$$

where c is a point between a and x, and interchanging a and x, that

$$g(a) = g(x) + (a-x)\,g'(x) + \frac{1}{2}(a-x)^2 g''(c'), \tag{11}$$

where c' is a point between a and x. Substituting (11) into the expression for $E_1'(g, x)$, we get

$$E_1'(g, x) = -\frac{1}{3}\big(2g(x) + g(a)\big)\frac{1}{2}(x-a)^2\, g''(c'),$$

and so with M_0 a bound for g on I and M_2 a bound for g'' on I, we have

$$|E_1'(g, x)| \leq \frac{1}{2} M_0 M_2 (x-a)^2, \quad x \in I.$$

Whence, since $E_1(g, a) = 0$, integrating the above inequality on I yields

$$|E_1(g, x)| \leq \frac{1}{6} M_0 M_2 (x-a)^3, \quad a \leq x \leq b. \tag{12}$$

To improve the approximation by the cone sums, we consider the family Π_e of partitions of I consisting of n equal length subintervals of I and apply the estimate (12) on each subinterval. Let then $\mathcal{P}_n = \{I_k^n\}$ be a partition of $I = [a, b]$

in Π_e, where $I_k^n = [x_{k-1}^n, x_k^n]$, and $x_k^n = a + (b-a)k/n$, for $k = 1, \ldots, n$. Then $(x_k^n - x_{k-1}^n) = (b-a)/n$, and from (12), it follows that

$$\left| \int_{I_k^n} g^2 - \frac{(b-a)}{n} \frac{\left(g^2(x_k^n) + g^2(x_{k-1}^n) + g(x_k^n)g(x_{k-1}^n)\right)}{3} \right|$$

$$\leq \frac{1}{6} M_0 M_2 \frac{(b-a)^3}{n^3},$$

and, since the absolute value of the sum is bounded by the sum of the absolute values, summing over k, the n above inequalities yields

$$\left| \int_I g^2 - C_n(g^2) \right| \leq \frac{1}{6} M_0 M_2 \frac{(b-a)^3}{n^2},$$

as we set out to prove.

Note that when $f(x) = (mx + b)^2$, with $m, b \geq 0$, $M_2 = 0$, and so the error is 0.   □

The cone expression $C_n(g^2)$ is a familiar one because, except for a factor of π, it corresponds to the calculation of the volume of a solid of revolution by the frustum method [52]. In this case, the disk method and the frustum method yield the same result. However, this is not the case when computing surface areas. The approximating expression by the frustum method is given by

$$A_n(f) = \frac{(b-a)}{n} \sum_{k=1}^{n} \pi \left(f(x_{k-1}) + f(x_k)\right) \sqrt{1 + f'(c_k)^2},$$

where $c_k \in (x_{k-1}, x_k)$, and by Bliss' theorem,

$$\lim_n A_n(f) = 2\pi \int_{[a,b]} f \sqrt{1 + f'^2},$$

which is the correct result. The failure of the disk method in this context is discussed in [52].

2.2 Rate of Approximation

We will examine next the rate of the approximation to $\int_I f$ when the Riemann sums of f consist of the averages of the values of f evaluated at n equally spaced points through I, under various conditions of the function integrated. We will begin by restricting ourselves to $I = [0, 1]$ and consider the admissible family Π_e of partitions of I. These sums include the one endpoint averages, such as the left and right Riemann sums of f. Specifically, let $C_L^n = \{0, 1/n, \ldots, (n-1)/n\}$ and $C_R^n = \{1/n, 2/n, \ldots, 1\}$ denote the tags corresponding to the left endpoints and the right endpoints of the intervals in the partitions $\mathcal{P}_n$ of I in Π_e, respectively. For simplicity, we will denote the left Riemann sum by $S_L(f, \mathcal{P}_n)$,

$$S_L(f, \mathcal{P}_n) = S_{\Pi_e}(f, \mathcal{P}_n, C_L^n) = \frac{1}{n} \sum_{k=0}^{n-1} f(k/n),$$

and the right Riemann sum by $S_R(f, \mathcal{P}_n)$,

$$S_R(f, \mathcal{P}_n) = S_{\Pi_e}(f, \mathcal{P}_n, C_R^n) = \frac{1}{n} \sum_{k=1}^{n} f(k/n).$$

Note that if f is increasing, $S_L(f, \mathcal{P}_n) = L_{\Pi_e}(f, \mathcal{P}_n)$ and $S_R(f, \mathcal{P}_n) = U_{\Pi_e}(f, \mathcal{P}_n)$, and vice versa for decreasing f.

We are then interested in the quantities

$$\Delta_{R,n}(f) = \int_I f - S_R(f, \mathcal{P}_n), \quad \Delta_{L,n}(f) = \int_I f - S_L(f, \mathcal{P}_n), \quad n = 1, 2, \ldots$$

The sequences $\{\Delta_{R,n}(f)\}$, $\{\Delta_{L,n}(f)\}$ are called the *Riemann coefficients* of f, and their behaviour resembles that of the Fourier coefficients of f, [19]. Just as the Fourier coefficients of f, the Riemann coefficients of f tend to 0 as $n \to \infty$, and for arbitrary Riemann integrable functions, the convergence is arbitrarily slow. Indeed, we have:

Proposition 4 *Let $\{\varepsilon_n\}$ be a positive sequence that decreases monotonically to* 0. *Then there is a Riemann integrable function f on $I = [0, 1]$ such that $|\Delta_{R,n}(f)| \geq \varepsilon_n$, for all n.*

Proof Let

$$f(x) = \begin{cases} 0, & x = 0, \text{ or } x \text{ irrational in } (0, 1), \\ \varepsilon_q, & x = p/q \text{ rational in } (0, 1], p, q \text{ relatively prime}. \end{cases}$$

First observe that f is continuous at each irrational x in I. Indeed, given $\varepsilon > 0$, let q be large enough so that $\varepsilon_q \leq \varepsilon$, let $\eta = \min\{|x - h/m| : 1 \leq h \leq m, 1 \leq m \leq q\}/2$, and pick $\delta > 0$ so that $(x - \delta, x + \delta) \subset (x - \eta, x + \eta) \cap (0, 1)$. Let then $y \in (x - \delta, x + \delta)$. If y is irrational, we have $|f(x) - f(y)| = 0 < \varepsilon$, and for rational $y = p'/q'$, since $q < q'$, we have $|f(x) - f(y)| = \varepsilon_{q'} \leq \varepsilon_q \leq \varepsilon$, and the continuity obtains.

To verify that f is Riemann integrable on I, we will invoke (4) above. Since the irrationals are dense in I, $L(f, \mathcal{P}) = 0$ for all partitions $\mathcal{P}$ of I, so, given $\varepsilon > 0$, it suffices to produce a partition $\mathcal{P}$ of I such that $U(f, \mathcal{P}) \leq \varepsilon$. Pick q

large enough so that $\varepsilon_q \leq \varepsilon/2$, and observe that there are a finite number N of distinct rational numbers p'/q' in I with $q' < q$, a rough estimate being $N < q^2$. Because the number of points is finite, we are able to pick pairwise disjoint intervals $I_k = [x_{k,l}, x_{k,r}]$, $1 \leq k \leq N$ each containing a different rational point and $|I_k| \leq \varepsilon/2N$. Let $\{J_h\}$ be the collection of subintervals of I defined as $J_0 = [0, x_{1,l}]$, $J_1 = [x_{1,r}, x_{2,l}], \ldots, J_{N-1} = [x_{N-1,r}, x_{N,l}]$, $J_N = [x_{N,r}, 1]$; J_0 or J_N may be empty. Then $\mathcal{P} = \{I_k\} \cup \{J_h\}$ is a partition of I, and with $M_k = \sup_{I_k} f$ and $M'_h = \sup_{J_h} f$, we will calculate

$$U(f, \mathcal{P}) = \sum_{k=1}^{N} M_k |I_k| + \sum_{h=0}^{N} M'_h |J_h| = A + B, \tag{13}$$

say.

Note that since $f(x) \leq 1$ throughout I, $M_k \leq 1$ for all k, and

$$A \leq \sum_{k=1}^{N} |I_k| \leq \frac{\varepsilon}{2N} N = \frac{\varepsilon}{2}.$$

Also, since each J_h contains only rationals p'/q' with $q' > q$, $M'_h \leq \varepsilon_q \leq \varepsilon/2$ for all h, and since the J_h are pairwise disjoint,

$$B \leq \frac{\varepsilon}{2} \sum_{h=0}^{N} |J_h| \leq \frac{\varepsilon}{2} |I| = \frac{\varepsilon}{2}.$$

Thus putting these estimates in (13), it follows that $U(f, \mathcal{P}) \leq \varepsilon/2 + \varepsilon/2 = \varepsilon$, and f is Riemann integrable by (4). Moreover, since $L(f) = 0$, $\int_I f = 0$.

Finally, if $k/n = p_{k,n}/q_{k,n}$ with $p_{k,n}, q_{k,n}$ relatively prime, then $q_{k,n} \leq n$ and $\varepsilon_n \leq \varepsilon_{q_{k,n}}$ for each $k = 1, \ldots, n$, and $n = 1, 2 \ldots$ Whence

$$|\Delta_{R,n}(f)| = \left| \int_I f - S_R(f, \mathcal{P}_n) \right| = S_R(f, \mathcal{P}_n)$$

$$= \frac{1}{n} \sum_{k=1}^{n} f(k/n) = \frac{1}{n} \sum_{k=1}^{n} \varepsilon_{q_{k,n}} \geq \varepsilon_n,$$

as we set out to prove.   □

The function f in Proposition 4 is a variant of the Thomae's function (introduced in 1875), which assumes the value 0 if x is an irrational in $(0, 1)$ and $1/q$ if $x = p/q$ where p, q are relatively prime. A caveat: a 21 year old Vito Volterra showed that there is no function defined on I that is continuous at the rationals and discontinuous at the irrationals of I, [104]. On the other hand, as we have seen, the characteristic function χ_C of the Cantor set, which has uncountably many discontinuities, is

Riemann integrable. And the Dirichlet function, which assumes the value 0 if x is rational and 1 if x is irrational in $[0, 1]$, satisfies $0 = L(f, \mathcal{P}) < U(f, \mathcal{P}) = 1$ for all partitions $\mathcal{P}$ of I and fails to be integrable on $[0, 1]$.

It is not apparent that Proposition 4 can be improved to obtain a simultaneous rate of convergence for all continuous functions. In fact, things get barely better for continuous functions. We will consider functions represented by a Fourier cosine series and will make use of the familiar Lagrange trigonometric identity,

$$\sum_{k=1}^{N} \cos(2kx) = \frac{\sin(Nx)\cos((N+1)x)}{\sin(x)}, \quad \text{if } \sin(x) \neq 0, \tag{14}$$

and $= N$ otherwise.

For these functions, $\Delta_{R,n}(f)$ can be computed exactly. Indeed, we have:

Proposition 5 *Let f be the continuous function represented in the interval $I = [0, 1]$ by the absolutely convergent series $f(x) = \sum_{n=1}^{\infty} a_n \cos(2\pi nx)$, where $\sum_n |a_n| < \infty$. Then,*

$$\Delta_{R,N}(f) = -(a_N + a_{2N} + a_{3N} + \cdots), \quad N = 1, 2, \ldots$$

Proof First note that $\int_I f = 0$. Now, by the uniform convergence of the sum defining f, it readily follows that

$$\sum_{k=1}^{N} f(k/N) = \sum_{n=1}^{\infty} a_n \sum_{k=1}^{N} \cos(2\pi nk/N).$$

We will apply (14) with $x = \pi n/N$. Since $\sin(x) = 0$ for the multiples $n = kN$ of N, the sum is equal to N at those values. And, since for the other values of n, $\sin(N(\pi n)/N) = \sin(\pi n) = 0$, and $\sin(x) \neq 0$, also by (14) the sum is 0, and, consequently,

$$\sum_{n=1}^{\infty} a_n \sum_{k=1}^{N} \cos(2\pi nk/N) = N(a_N + a_{2N} + a_{3N} + \cdots),$$

and the conclusion follows. □

We will now define a continuous function whose Riemann sums converge arbitrarily slowly to its integral.

Proposition 6 *Let $\{\varepsilon_n\}$ be a positive sequence that decreases to 0. Then there is a continuous, periodic function f of period 1 defined on I such that*

$$\limsup_{n} \frac{|\Delta_{R,n}(f)|}{\varepsilon_n} = \infty. \tag{15}$$

Proof Let p_1 be a prime number such that $\varepsilon_{p_1} < 1/2$, and having picked prime numbers $p_1 < p_2 < \cdots < p_{k-1}$, let p_k be a prime number $> p_{k-1}$ such that $\varepsilon_{p_k} \leq 1/k2^k$. Let

$$f(x) = \sum_{k=1}^{\infty} \frac{1}{2^k} \cos(2\pi p_k x).$$

Then, by Proposition 5, $|\Delta_{R,p_k}(f)| = 2^{-k}$, all k. Hence,

$$\frac{|\Delta_{p_k}(f)|}{\varepsilon_{p_k}} \geq \frac{1}{2^k} k 2^k = k, \quad \text{all } k,$$

and, consequently, (15) holds.   □

We will discuss next some results for specific classes of functions. Recall that the *modulus of continuity* $\omega_f(\delta)$ of f on I is defined as

$$\omega_f(\delta) = \sup_{|x-y| \leq \delta} |f(x) - f(y)|, \quad x, y \in I.$$

Thus, $|f(x) - f(y)| \leq \omega_f(\delta)$ whenever $x, y \in I$ are such that $|x - y| \leq \delta$. Recall that uniformly continuous functions on I are characterized by the property that $\lim_{\delta \to 0+} \omega_f(\delta) = 0$.

We claim that for continuous functions f,

$$|\Delta_{R,n}(f)| \leq \omega_f(1/n).$$

The verification is immediate. Since

$$\Delta_{R,n}(f) = \int_I f - \frac{1}{n} \sum_{k=1}^{n} f(k/n) = \sum_{k=1}^{n} \int_{I_k^n} \big(f(x) - f(k/n)\big) dx, \tag{16}$$

and since $|f(x) - f(k/n)| \leq \omega_f(1/n)$ throughout I_k^n, each of the integrals on the right-hand side of (16) is bounded by

$$\int_{I_k^n} |f(x) - f(k/n)| \, dx \leq \frac{1}{n} \omega_f(1/n), \quad 1 \leq k \leq n,$$

and the conclusion follows by adding the n inequalities above.

A similar estimate holds for bounded monotone functions f defined on $I = [0, 1]$, which may be discontinuous at an infinite number of points. For instance, the increasing function f defined by $f(x) = 1 - 1/n$ if $1 - 1/n \leq x < 1 - 1/(n+1)$, for $n = 1, 2, \ldots$, and $f(1) = 1$, is discontinuous at every $x = 1 - 1/n$. Suppose then that f is monotone increasing in I; if f is decreasing, we argue with f replaced by $-f$. Let $\mathcal{P}_n = \{I_k^n\}$ be a partition of I in Π_e consisting of n intervals of equal length, all n. Since f is increasing, $m_k^n = \inf_{I_k^n} f = f((k-1)/n)$, and $M_k^n = \sup_{I_k^n} f = f(k/n)$. Hence,

$$U_{\Pi_e}(f, \mathcal{P}_n) - L_{\Pi_e}(f, \mathcal{P}_n)$$

$$= \frac{1}{n} \sum_{k=1}^{n} \left(f(k/n) - f((k-1)/n) \right) = \frac{1}{n} \left(f(1) - f(0) \right),$$

which tends to 0 as $n \to \infty$, and by (4), f is Riemann integrable on I. Moreover, since

$$\int_{I_k^n} \left| f(x) - f(k/n) \right| dx \le \frac{1}{n} \left(f(k/n) - f((k-1)/n) \right), \quad 1 \le k \le n,$$

summing the above n inequalities, it follows that

$$0 \le \Delta_{R,n}(f) \le \frac{1}{n} \left(f(1) - f(0) \right).$$

We note that the $O(1/n)$ estimate extends to the important class of functions of bounded variation, or BV functions, which can be expressed as a difference of monotone functions [46].

Then there are continuous functions that exhibit smoothness, namely, those that are Lipschitz on I. Recall that a function f defined on I is said to be *Lipschitz of order* α there, $0 < \alpha \le 1$, provided that for some constant L, which depends on f,

$$\left| f(x) - f(y) \right| \le L |x - y|^\alpha, \quad \text{all } x, y \in I.$$

For these functions, since

$$\left| f(x) - f(y) \right| \le L \frac{1}{n^\alpha}, \quad \text{all } x, y \in I_k^n, \tag{17}$$

from (16) and (17), it follows that the Riemann coefficients of f satisfy

$$\left| \Delta_{R,n}(f) \right| \le L \frac{1}{n^\alpha} \sum_{k=1}^{n} |I_k^n| = L \frac{1}{n^\alpha},$$

which is then the rate of approximation, and which, as we shall see below, is also the rate of decay of the Fourier coefficients of functions in the class Lipschitz α.

Finally, an observation of general nature that we will use freely in what follows: if f is Riemann integrable on I, then so are $f^+ = \max(f, 0)$, $f^- = \max(-f, 0)$, and $|f| = f^+ + f^-$. Indeed, $f^+, f^- \ge 0$, and

$$\text{osc}(f^+, I_k^n), \ \text{osc}(f^-, I_k^n) \le \text{osc}(f, I_k^n), \quad \text{all } k, n.$$

Thus,

$$\sum_{k=1}^{n} \mathrm{osc}\,(f^+, I_k^n)\,|I_k^n|, \ \sum_{k=1}^{n} \mathrm{osc}\,(f^-, I_k^n)\,|I_k^n| \leq \sum_{k=1}^{n} \mathrm{osc}\,(f, I_k^n)\,|I_k^n|,$$

and by (6), f^+, f^- are Riemann integrable. Then, so is $|f| = f^+ + f^-$.

2.3 Quadrature Rules

When a definite integral cannot be evaluated exactly, we appeal to approximation methods that, as we have seen in the case of the cone sums, can be complicated. Any method used to approximate a definite integral is called a *quadrature rule*. Historically, a quadrature is any process used to construct a square equal in area to that of some given figure. The basic idea of a quadrature rule is to replace the integral by a Riemann sum of the integrand evaluated at carefully selected tags, called *quadrature points* or *nodes*, multiplied by a number, known as the *quadrature weights*. We denote this by

$$\int_{[a,b]} f \approx \sum_{k=1}^{m} w_k f(x_k)$$

and call the above sum a *quadrature formula*. Informally, we approximate the integral of f by a weighted sum of function values and speak then of numerical integration or *numerical quadrature*. We always assume that the sum for $f = \chi_I$ is exact; in other words, $\sum_{k=1}^{n} w_k = (b - a)$.

Note that a quadrature rule may be transferred between intervals. Suppose, e.g., that the rule is written for $I = [0, 1]$, and so $0 \leq x_1 < x_2 < \ldots < x_m \leq 1$. To transfer the rule to the interval $[a, b]$, consider the transformation $y = a + (b - a)x$, which maps I onto $[a, b]$. The nodes x_k then become $a + (b - a)x_k$, the weights w_k become $(b - a)w_k$, $1 \leq k \leq m$, and the rule becomes

$$\int_{[a,b]} f \approx (b - a) \sum_{k=1}^{m} w_k f(a + (b - a)x_k).$$

Cruz Uribe and Neugebauer note that typically estimates for the quadrature rules are derived using polynomial interpolation, which leads to higher order derivatives of the function involved on the right-hand side of the estimate. However, the assumption that f has continuous higher order derivatives precludes us from estimating the error when approximating the integral of well-behaved functions such as $f(x) = \sqrt{x}$ on $[0, 1]$, see [22, 23, 34].

Now, one way to approximate the value of a definite integral is by the use of the endpoint quadrature rules. There are *open type* rules, which do not use the left endpoint, the right endpoint, or both endpoints, and *closed type* rules that incorporate the endpoints.

We will begin by estimating the rate of approximation of $\Delta_{R,n}(f)$ and $\Delta_{L,n}(f)$ and the right endpoint and the left endpoint rule, respectively. Since for these rules the error is 0 when f is constant, we might expect the error term to involve the first derivative f' of f. So, fix an interval $I = [a, b]$, and consider the simple right Riemann sum

$$\int_{[a,b]} f \approx (b - a) f(b),$$

and the associated error expression

$$E(f, x) = \int_{[a,x]} f - (x - a) f(x), \quad a \le x \le b.$$

The following argument to estimate $E(f, b)$ is found all too often in the literature. By the mean value theorem, $f(x) = f(b) + (x - b) f'(c)$ for some $c \in (x, b)$, and, consequently, $E(f, b)$ is equal to

$$\int_{[a,b]} f - (b - a) f(b) = \int_{[a,b]} \big(f(x) - f(b)\big) \, dx = \int_{[a,b]} (x - b) f'(c) \, dx$$

$$= f'(c) \int_{[a,b]} (x - b) \, dx = -\frac{(b - a)^2}{2} f'(c).$$

Although the conclusion may be correct, the argument is flawed as it fails to take into account that $c = c(x)$ depends on x. Also to the point, it is not apparent that $f'(c(x))$ is Riemann integrable on I. Indeed, if φ and ψ are Riemann integrable, then $\varphi \circ \psi$ is Riemann integrable provided φ is continuous, but not necessarily if φ is merely integrable. To see this when φ is merely integrable, let $\varphi(x) = 1$ if $x \ne 0$, and $\varphi(x) = 0$ if $x = 0$. Then with $\psi = f$ as in Proposition 4, $\varphi \circ \psi = \chi_Q$, the characteristic function of the rationals in I, which, since $0 = L(\chi_Q) \ne U(\chi_Q) = 1$, is not integrable. A more involved argument shows that there are continuous functions ψ so that the composition is not integrable [65].

Returning to $E(f, x)$, since $E'(f, x) = -(x - a) f'(x)$, with M_1 a bound for f' on I, it follows that $|E'(f, x)| \le M_1 |x - a|$, and since $E(f, a) = 0$, we have

$$|E(f, x)| \le \int_{[a,x]} |t - a| \, |f'(t)| \, dt \le M_1 (x - a)^2 / 2, \quad a \le x \le b.$$

This gives a bound for $E(f, x)$ for the simple right Riemann sums. And we can also compute the error exactly. Indeed, by the mean value theorem, it readily follows that for some $a \leq c \leq x$, with c depending on x,

$$E(f, x) = (x - a)E'(f, c) = -(x - a)(c - a)f'(c), \quad a \leq x \leq b.$$

As with the cone sums, to improve the computation and estimation of the error term, we consider the *compound*, or *composite*, right Riemann sum of f, which arises when the interval of integration is subdivided into a number of subintervals and the right Riemann sum, or in general a fixed rule of integration, is applied to each of the subintervals.

Note that for Riemann integrable functions f, the composite quadrature formulas converge to $\int_I f$, [25]. To see this, let $I = [a, b]$, and with $0 \leq t_1 < \ldots < t_m \leq 1$, let

$$\int_{[a,b]} f \approx (b - a) \sum_{k=1}^{m} w_k f(a + (b - a)t_k), \quad \sum_{k=1}^{m} w_k = 1.$$

Let $\mathcal{P}_n = \{I_h^n\}$ be a partition of I in Π_e, $I_h = [x_{h-1}, x_h]$, $1 \leq h \leq n$. Then, for each J_h, by the transfer rule,

$$\int_{J_h} f \approx (x_h - x_{h-1}) \sum_{k=1}^{m} w_k f(x_{h-1} + (x_h - x_{h-1})t_k),$$

$$= \frac{(b - a)}{n} \sum_{k=1}^{m} w_k f(x_{h-1} + ((b - a)/n)t_k),$$

and, therefore, summing,

$$\int_{[a,b]} f \approx \sum_{h=1}^{n} \frac{(b - a)}{n} \sum_{k=1}^{m} w_k f(x_{h-1} + ((b - a)/n)t_k)$$

$$= \sum_{k=1}^{m} w_k \frac{(b - a)}{n} \sum_{h=1}^{n} f(x_{h-1} + ((b - a)/n)t_k).$$

Now, for each k, $C^k = \{x_{h-1} + ((b - a)/n) t_k : 1 \leq h \leq n\}$ is a set of tags for the partition $\mathcal{P}_n = \{I_h^n\}$ of I, and consequently, the innermost sum above is a Riemann sum for f with limit $\int_I f$ as $n \to \infty$. Whence, since $\sum_{k=1}^{m} w_k = 1$, the limit as $n \to \infty$ of the expression is $\int_I f$.

Back to the computations, we will make use of the intermediate value property for a positive linear combination of derivatives [93]. More precisely:

General Darboux Property *Let f be a function defined on $I = [a, b]$ that is differentiable on (a, b). Let λ_k be positive real numbers and $c_k \in (a, b)$, for*

$1 \leq k \leq n$. Then there is $c \in (a, b)$ such that

$$\sum_{k=1}^{n} \lambda_k f'(c_k) = f'(c) \sum_{k=1}^{n} \lambda_k.$$

Proof Suppose first that $\sum_{k=1}^{n} \lambda_k = 1$. If all the $f'(c_k)$ are equal, the assertion is trivial. So, let c_m, c_M be such that $\min_k f'(c_k) = f'(c_m) < f'(c_M) = \max_k f'(c_k)$, and observe that

$$f'(c_m) < \sum_{k=1}^{n} \lambda_k f'(c_k) < f'(c_M).$$

Now, by Darboux's theorem for derivatives, there exists $c \in (c_m, c_M)$ such that $f'(c) = \sum_{k=1}^{n} \lambda_k f'(c_k)$, and the conclusion holds in this case. In the general case, let $\mu_h = \lambda_h / \sum_{k=1}^{n} \lambda_k$, $1 \leq h \leq n$, and observe that

$$\sum_{k=1}^{n} \lambda_k f'(c_k) = \sum_{k=1}^{n} \lambda_k \sum_{h=1}^{n} \mu_h f'(c_h),$$

which since $\sum_{h=1}^{n} \mu_h = 1$ by the first part of the proof implies that

$$\sum_{k=1}^{n} \lambda_k f'(c_k) = f'(c) \sum_{k=1}^{n} \lambda_k,$$

as we set out to prove.   □

We will now consider the composite right Riemann sums of f. We divide I into n equal length subintervals, apply the error identity, and estimate on each subinterval of I. We will carry out the argument for the exact expression of the error, the other being similar. So, let $\mathcal{P}_n = \{I_k^n\}$ be a partition of $I = [a, b]$ in Π_e, where $I_k^n = [x_{k-1}^n, x_k^n]$, and $x_k^n = a + (b-a)k/n$, for $k = 1, \ldots, n$. Then $(x_k^n - x_{k-1}^n) = (b-a)/n$, and from the above identity, it follows that on each I_k^n,

$$\int_{I_k^n} f - (x_k^m - x_{k-1}^n) f(x_k^n) = -\frac{(b-a)}{n}(c_k^n - x_{k-1}^n)f'(c_k^n), \quad 1 \leq k \leq n,$$

and, consequently, summing the above n identities, it follows that

$$\Delta_{R,n}(f) = -\frac{(b-a)}{n} \sum_{k=1}^{n} (c_k^n - x_{k-1}^n)f'(c_k^n).$$

Therefore, by the general Darboux property, an exact expression for the error is

$$\Delta_{R,n}(f) = -\frac{(b-a)}{n}\left(\sum_{k=1}^{n}(c_k^n - x_{k-1}^n)\right)f'(c),$$

where $c_k^n \in I_k^n$ and $c \in (a, b)$.

A careful argument also gives the asymptotic behaviour of $n\Delta_{R,n}(f)$ as $n \to \infty$. Indeed, we have:

Theorem 2 *Let f be a differentiable function defined on $I = [a, b]$ such that its derivative f' is Riemann integrable on I. Then,*

$$\lim_{n} n\Delta_{R,n}(f) = -\frac{(b-a)}{2}\left(f(b)-f(a)\right), \lim_{n} n\Delta_{L,n}(f) = \frac{(b-a)}{2}\left(f(b)-f(a)\right).$$

Proof Assume first that $I = [0, 1]$. Referring to (16), let $\mathcal{P}_n = \{I_k^n\}$ be a partition of I in Π_e. Then, for $1 \leq k \leq n$, by the mean value theorem, for $x \in I_k^n$, there exists $c_k^n \in I_k^n$ that depends on x, k, and n, such that

$$f(x) - f(k/n) = (x - k/n)f'(c_k^n),$$

and, consequently, since $(x - k/n) \leq 0$ throughout I_k^n, with $m_k^{n'} = \inf_{I_k^n} f'$ and $M_k^{n'} = \sup_{I_k^n} f'$, we have

$$M_k^{n'}(x - (k/n)) \leq f(x) - f(k/n) \leq (x - (k/n))m_k^n.$$

Hence, since $\int_{I_k^n}(x - (k/n))\,dx = -1/2n^2$, integrating over I_k^n and summing the n inequalities above, it follows that

$$-\frac{1}{2}U_{\Pi_e}(f', \mathcal{P}_n) \leq n\,\Delta_{R,n}(f) \leq -\frac{1}{2}L_{\Pi_e}(f', \mathcal{P}_n).$$

Now, since $\lim_{n} L_{\Pi_e}(f', \mathcal{P}_n) = \lim_{n} U_{\Pi_e}(f', \mathcal{P}_n) = \int_{I} f'$, it readily follows that

$$\lim_{n} n\,\Delta_{R,n}(f) = -\frac{1}{2}\int_{I} f' = -\frac{f(1) - f(0)}{2},$$

which is what we set out to prove in this case.

Also, since $n\,S_L(f, \mathcal{P}_n) = n\,S_R(f, \mathcal{P}_n) - \left(f(1) - f(0)\right)$, it follows that

$$\lim_{n} n\,\Delta_{L,n}(f) = \lim_{n} n\Delta_{L,n}(f) - \left(f(1) - f(0)\right) = \frac{f(1) - f(0)}{2},$$

and the proof is finished in this case.

When f is defined on an arbitrary interval $[a, b]$, we appeal to a scaling argument. If f is defined on $[a, b]$ and has a Riemann integrable derivative f' there, put

$$g(x) = f\big((b - a)x + a\big), \quad x \in I = [0, 1].$$

Then g is a differentiable function on $I = [0, 1]$ that verifies

$$\int_I g = \frac{1}{b - a} \int_{[a,b]} f, \quad g(1) = f(b), \quad g(0) = f(a),$$

and

$$\frac{1}{n} \sum_{k=1}^{n} g(k/n) = \frac{1}{n} \sum_{k=1}^{n} f(a + (b - a)(k/n)).$$

Therefore, multiplying through by $(b - a)$, the first part of the argument applied to g yields that

$$\lim_n n \left(\int_{[a,b]} f - \frac{b - a}{n} \sum_{k=1}^{n} f\big(a + (b - a)(k/n)\big) \right) = -(b - a)\frac{f(b) - f(a)}{2}.$$

For the left Riemann sums of f, we proceed as above and the proof is finished.

□

A couple of observations concerning Theorem 2. First, the knowledge of the exact rate of approximation can be used to advantage. Suppose that for a differentiable function f on I with a Riemann integrable derivative f' and integers n_1, n_2, we are interested in evaluating

$$\lim_n \left(\sum_{k=1}^{n+n_1} f(k/(n + n_1)) - \sum_{k=1}^{n+n_2} f(k/(n + n_2)) \right).$$

Since the above expression can be written as

$$(n + n_1) S_R(f, \mathcal{P}_{n+n_1}) - (n + n_2) S_R(f, \mathcal{P}_{n+n_2})$$

$$= (n + n_1)\left(S_R(f, \mathcal{P}_{n+n_1}) - \int_I f \right) - (n + n_2)\left(S_R(f, \mathcal{P}_{n+n_2}) - \int_I f \right)$$

$$+ (n_1 - n_2) \int_I f,$$

by Theorem 2, the limit as $n \to \infty$ is equal to $(n_1 - n_2) \int_I f$.

Also, an examination of Theorem 2 suggests that since the rate of approximation of $n\Delta_{R,n}(f)$ is equal to that of $n\Delta_{L,n}(f)$ and of opposite sign, the linear combination

$$\frac{1}{2}\big(S_L(f,\mathcal{P}_n) + S_R(f,\mathcal{P}_n)\big),$$

should be a better approximation. This is indeed the case, and we will discuss it below when considering the trapezoidal rule.

It is also of interest to deal with the *uneven tags* that result by choosing tag points at a fixed distance from one of the endpoints in each I_k^n. Specifically, for $0 < \alpha < 1$, let $C_\alpha^n = \{(1-\alpha)/n, (2-\alpha)/n, \ldots, (n-\alpha)/n\}$, and consider the Riemann sum

$$S_{U_\alpha}(f,\mathcal{P}_n) = S_{\Pi_e}(f,\mathcal{P}_n, C_\alpha^n) = \frac{1}{n}\sum_{k=1}^{n} f\Big(\frac{k-\alpha}{n}\Big),$$

and the associated error terms

$$\Delta_{U_\alpha,n}(f) = \int_I f - S_{U_\alpha}(f,\mathcal{P}_n).$$

We will prove the result as it applies to the product of functions. Specifically, given a finite collection of Riemann integrable functions $f_1, \ldots, f_N$ defined on I and a multi-index $\alpha = (\alpha_1, \ldots, \alpha_N)$ with $0 \le \alpha_j < 1$ for $1 \le j \le N$, let

$$\Delta_{U_\alpha,n}\Big(\prod_{i=1}^{N} f_i\Big) = \int_I \prod_{i=1}^{N} f_i - \frac{1}{n}\sum_{k=1}^{n}\prod_{i=1}^{N} f_i\Big(\frac{(k-\alpha_i)}{n}\Big).$$

We then have:

Theorem 3 *Let $f_1, \ldots, f_N$ be a finite collection of differentiable functions defined on I, with Riemann integrable derivatives f_j' for $1 \le j \le N$. Then we have*

$$\lim_n n\,\Delta_{U_\alpha,n}\Big(\prod_{i=1}^{N} f_i\Big) = \frac{1}{2}\Big(\prod_{i=1}^{N} f_i(0) - \prod_{i=1}^{N} f_i(1)\Big) + \sum_{j=1}^{N}\alpha_j\int_I\Big(f_j'\prod_{i\ne j,i=1}^{N} f_i\Big).$$

Proof We begin by writing

$$n\,\Delta_{U_\alpha,n}\Big(\prod_{i=1}^{N} f_i\Big) = n\Big(\int_I \prod_{i=1}^{N} f_i - \frac{1}{n}\sum_{k=1}^{n}\prod_{i=1}^{N} f_i\Big(\frac{k}{n}\Big)\Big)$$

$$+ \sum_{k=1}^{n}\Big(\prod_{i=1}^{N} f_i\Big(\frac{k}{n}\Big) - \prod_{i=1}^{N} f_i\Big(\frac{(k-\alpha_i)}{n}\Big)\Big). \tag{18}$$

Concerning the second summand in (18), a straightforward computation shows that for each k we have

$$\prod_{i=1}^{N} f_i\left(\frac{k}{n}\right) - \prod_{i=1}^{N} f_i\left(\frac{(k-\alpha_i)}{n}\right) = \sum_{j=1}^{N} \left(f_j\left(\frac{k}{n}\right) - f_j\left(\frac{(k-\alpha_j)}{n}\right)\right) \prod_{i\neq j, i=1}^{N} f_i\left(x_k^{i,n}\right),$$

where $x_k^{i,n} = (k-\alpha_i)/n$ for $1 \leq i < j$, and $x_k^{i,n} = k/n$ for $j < i \leq N$.

Now, by the mean value theorem, there are $c_k^{j,n}$ between $(k-\alpha_j)/n$ and k/n, such that

$$f_j\left(\frac{k}{n}\right) - f_j\left(\frac{(k-\alpha_j)}{n}\right) = \frac{\alpha_j}{n} f_j'(c_k^{j,n}), \quad 1 \leq j \leq N,$$

and, consequently, the second summand in (18) above becomes

$$\sum_{j=1}^{N} \alpha_j \frac{1}{n} \sum_{k=1}^{n} f_j'(c_k^{j,n}) \prod_{i\neq j, i=1}^{N} f_i\left(x_k^{i,n}\right),$$

which, by Bliss' theorem, satisfies

$$\lim_n \sum_{j=1}^{N} \alpha_j \frac{1}{n} \sum_{k=1}^{n} f_j'(c_k^{j,n}) \prod_{i\neq j, i=1}^{N} f_i\left(x_k^{i,n}\right) = \sum_{j=1}^{N} \alpha_j \int_I \left(f_j' \prod_{i\neq j, i=1}^{N} f_i\right). \tag{19}$$

Also, since the first summand in (18) is equal to $n \, \Delta_{R,n}(\prod_{i=1}^{N} f_i)$, by Theorem 2, it follows that

$$\lim_n n \, \Delta_{R,n}\left(\prod_{i=1}^{N} f_i\right) = \frac{1}{2} \left(\prod_{i=1}^{N} f_i(0) - \prod_{i=1}^{N} f_i(1)\right),$$

which combined with (19) gives the desired conclusion. □

The case $N = 1$ is a most interesting instance of Theorem 3. With $f_1 = f$, the conclusion reads

$$\lim_n n \, \Delta_{U_\alpha, n}(f) = \frac{f(0) - f(1)}{2} + \alpha \int_I f' = \left(\alpha - \frac{1}{2}\right)\left(f(1) - f(0)\right),$$

and, consequently, the choice $\alpha = 1/2$ gives the best approximation, with the order of approximation of the error term strictly smaller than $1/n$.

To clarify the situation, we will consider the partitions where the tags are the midpoints of the partition intervals. So, let m_k^n denote the midpoint of the intervals I_k^n, i.e., $m_k^n = (k-1/2)/n$, $1 \leq k \leq n$, and consider the Riemann sums with tags

$$C_M^n = \{m_1^n, \ldots, m_n^n\},$$

$$S_M(f, \mathcal{P}_n) = S_{\Pi_e}(f, \mathcal{P}_n, C_M^n) = \frac{1}{n} \sum_{k=1}^{n} f(m_k^n),$$

and the associated error term

$$\Delta_{M,n}(f) = \int_I f - S_M(f, \mathcal{P}_n).$$

Since $\Delta_{M,n}(f) = 0$ if f is linear but not if $f(x) = x^2$, we might expect that an error term should involve a bound for f''. On $I = [a, b]$, consider

$$E(f, x) = \int_{[a,x]} f - (x - a)f\big((a + x)/2\big), \quad a \leq x \leq b.$$

Then by the mean value theorem with some c between x and $(a + x)/2$, we have

$$E'(f, x) = f(x) - f\big((a + x)/2\big) - \frac{(x - a)}{2} f'\big((a + x)/2\big)$$
$$= \frac{(x - a)}{2}\big(f'(c) - f'((a + x)/2)\big),$$

and again by the mean value theorem with c' between c and $(a + x)/2$,

$$E'(f, x) = \frac{(x - a)}{2}\Big(c - \big((a + x)/2\big)\Big) f''(c'),$$

and so, with M_2 a bound for f'', it follows that $|E'(f, x)| \leq (M_2/4)(x - a)^2$, which, since $E(f, a) = 0$, yields

$$|E(f, x)| \leq \frac{1}{12} M_2 (x - a)^3, \quad a < x \leq b.$$

As for the asymptotic estimate, we will prove [75]:

Theorem 4 *Let f be a function defined on $I = [a, b]$ that is twice differentiable there, with f'' Riemann integrable on I. Then we have*

$$\lim_n n^2 \Delta_{M,n}(f) = (b - a)^2 \frac{f'(b) - f'(a)}{24}.$$

Proof We will consider $I = [0, 1]$, for then the proof may be completed by scaling. Let $\mathcal{P}_n = \{I_k^n\}$ be a partition of I in Π_e, and write

$$\Delta_{M,n}(f) = \sum_{k=1}^{n} \int_{I_k^n} \left(f(x) - f(m_k^n) \right) dx.$$

For $1 \leq k \leq n$, suppose that $x \in I_k^n$. Then by the mean value theorem for the second derivative [69], we have

$$f(x) = f(m_k^n) + (x - m_k^n) f'(m_k^n) + \frac{1}{2}(x - m_k^n)^2 f''(\zeta_k^n),$$

where ζ_k^n is some number between x and m_k^n that depends on x, k, and n, and, consequently, since $(x - k/n)^2 \geq 0$ for $x \in I_k^n$, with $m_k^{n''} = \inf_{I_k^n} f''$ and $M_k^{n''} = \sup_{I_k^n} f''$, we have

$$m_k^{n''} \frac{1}{2}(x - m_k^n)^2 \leq f(x) - f(m_k^n) - (x - m_k^n) f'(m_k^n) \leq \frac{1}{2}(x - m_k^n)^2 M_k^{n''}.$$

Hence, since $\int_{I_k^n}(x - m_k^n) \, dx = 0$ and $\int_{I_k^n}(x - m_k^n)^2 dx = 1/12n^3$, integrating over I_k^n and summing the above n inequalities, it follows that

$$\frac{1}{2}\frac{1}{12n^2} L_{\Pi_e}(f'', \mathcal{P}_n) \leq \Delta_{M,n}(f) \leq \frac{1}{2}\frac{1}{12n^2} U_{\Pi_e}(f'', \mathcal{P}_n).$$

Now, $\lim_n L_{\Pi_e}(f'', \mathcal{P}_n) = \lim_n U_{\Pi_e}(f'', \mathcal{P}_n) = \int_I f''$, and we conclude that

$$\lim_n n^2 \Delta_{M,n}(f) = \frac{1}{24} \int_I f'' = \frac{f'(1) - f'(0)}{24},$$

and the result has been established.   □

As the comment following Theorem 2 suggests, we consider the linear combination

$$T(f, \mathcal{P}_n) = \frac{1}{2}\left(S_L(f, \mathcal{P}_n) + S_R(f, \mathcal{P}_n) \right),$$

with the expectation to improve on the approximation. Now, on $I = [0, 1]$, with $x_h^m = h/m$ for $1 \leq h \leq n, n = 1, 2, \ldots$, we have

$$T(f, \mathcal{P}_n) = \frac{1}{n} \sum_{k=1}^{n} \frac{f(x_{k-1}^n) + f(x_k^n)}{2}, \quad n = 1, 2, \ldots$$

where the area above each interval I_k^n of the partition is approximated by a trapezoid with vertices $(x_{k-1}^n, 0)$, $(x_k^n, 0)$, $(x_{k-1}^n, f(x_{k-1}^n))$, and $(x_k^n, f(x_k^n))$. Hence, the name *trapezoid sum* for this expression.

We begin by considering the case of uneven length partition intervals; to emphasize this, we display the partition in the notation of the sums. Then, for general partitions $Q_n = \{J_k^n\}$, with $J_k^n = [x_{k,l}^n, x_{k,r}^n]$, $1 \leq k \leq m_n$, of partitions of I with $\|Q_n\| \to 0$, let

$$T(f, Q_n) = \frac{1}{2} \sum_{k=1}^{m_n} \left(f(x_{k,l}^n) + f(x_{k,r}^n) \right) |J_k^n|$$

and consider the corresponding error term

$$\Delta_{T,Q_n}(f) = \int_I f - T(f, Q_n), \quad n = 1, 2, \ldots$$

Since the above expression integrates a linear function exactly, but not so a quadratic, we might expect the error term for this inequality to involve the second derivative f'' of f. In this general setting, we have [31]:

Proposition 7 *Let $I = [a, b]$ be a closed bounded interval, and let f be a twice differentiable function defined On I. Then, if f'' is Riemann integrable on I, for partitions $Q_n = \{J_k^n\}$ of I, we have*

$$\left| \Delta_{T,Q_n}(f) \right| \leq A(f),$$

where $A(f)$ assumes one of these two forms,

$$\frac{1}{12} \left(\sup_I |f''| \right) \left(\sum_{k=1}^{m_n} |J_k^n|^3 \right), \quad \text{or} \quad \frac{1}{8} \left(\int_I |f''| \right) \left(\sum_{k=1}^{m_n} |J_k^n|^2 \right).$$

Proof We consider first the case when the partition consists of the single interval I. Then, since f is twice differentiable in I, for $x \in I$, we have

$$\left((b - x)(x - a) f'(x) \right)' = (b - x)(x - a) f''(x) - (2x - (a + b)) f'(x),$$

and, consequently, by the fundamental theorem of calculus and integration by parts,

$$\int_I (b - x)(x - a) f''(x) \, dx = \int_I \left(2x - (a + b) \right) f'(x) \, dx$$

$$= \left((2x - (a + b)) f(x) \right]_a^b - 2 \int_I f,$$

which can be rewritten as

$$\int_I f - (b - a) \frac{(f(b) + f(a))}{2} = -\frac{1}{2} \int_I (b - x)(x - a) f''(x) \, dx.$$

Hence,

$$\left| \Delta_{T,1}(f) \right| \leq \frac{1}{2} \int_I (b - x)(x - a) |f''(x)| \, dx,$$

which can in turn be estimated by

$$\max_I (b - x)(x - a) \frac{1}{2} \int_I |f''| = \frac{1}{8} \left(\int_I |f''| \right) (b - a)^2,$$

or by

$$\frac{1}{2} \sup_I |f''| \int_I (b - x)(x - a) \, dx = \frac{1}{12} \sup_I |f''| (b - a)^3. \tag{20}$$

Equation (20) yields the familiar trapezoid inequality. To improve the approximation, we will consider the composite trapezoid rule. We will complete the proof of (20) in this case.

Let $Q_n = \{J_1^n, \ldots, J_{m_n}^n\}$, where $J_k^n = [x_{k,l}^n, x_{k,r}^n]$, be a partition of I where $x_{1,l}^n = a$ and $x_{m,r}^n = b$. Then, since

$$\Delta_{T,Q_n}(f) = \sum_{k=1}^{m_n} \left(\int_{J_k^n} f - \frac{1}{2} \left(f(x_{k,l}^n) + f(x_{k,r}^n) \right) |J_k^n| \right)$$

from (20), it follows that

$$\left| \Delta_{T,Q_n}(f) \right| \leq \frac{1}{12} \sum_{k=1}^{m_n} \sup_{J_k^n} |f''| \, |J_k^n|^3 \leq \frac{1}{12} \sup_I |f''| \sum_{k=1}^{m_n} |J_k^n|^3.$$

The proof of the other statement follows along similar lines and is left to the reader. $\qquad\Box$

The usual trapezoid estimate does not apply to some simple functions, e.g., $f(x) = x^\alpha, 1 < \alpha < 2$, defined on $I = [0, 1]$. It then follows that $f''(x) = \alpha(\alpha - 1)x^{\alpha - 2} \to \infty$ as $x \to 0^+$, and no estimate with $\sup_I |f''|$ on the right-hand side of the inequality is available. On the other hand, $\int_{[\varepsilon, 1]} |f''| \leq \alpha(\alpha - 1)^2$ for $\varepsilon > 0$, and so

$$|\Delta_{T,1}(f)| \leq \limsup_{\varepsilon \to 0^+} \left| \int_{[\varepsilon, 1]} f - (1 - \varepsilon) \frac{f(1) + f(\varepsilon)}{2} \right|$$

$$\leq \limsup_{\varepsilon \to 0^+} \frac{1}{8} \int_{[\varepsilon,1]} |f''| \leq \frac{1}{8}\alpha(\alpha - 1)^2,$$

and this estimate is available.

Yet another way to estimate $A(f)$ is by Hölder's inequality applied to the expression that bounds $\Delta_{T,1}(f)$. In [31], the authors show by means of examples that of the three estimates available, no estimate is better than the other two. They also consider applications to estimates involving different means, including arithmetic, geometric, harmonic, p-logarithmic, and identric.

Restricting the partitions to Π_e, it is possible to obtain a precise asymptotic behaviour:

Theorem 5 *Let f be a twice differentiable function defined on $I = [a, b]$ such that f'' is Riemann integrable on I. Then*

$$\lim_n n^2 \Delta_{T,n}(f) = \frac{1}{12}(b - a)^2\big(f'(a) - f'(b)\big).$$

Proof By scaling, it suffices to prove the case $[a, b] = [0, 1]$. We consider the case $n = 1$ first. If f is twice differentiable in I, by a simple application of the mean value theorem for derivatives, it follows that for $a < x \leq b$, we have

$$f(x) = f(a) + (x - a)f'(a) + \frac{1}{2}(x - a)^2 f''(c),$$

where c is a point between a and x. In particular for $x = b$, we have

$$f(b) = f(a) + (b - a)f'(a) + \frac{1}{2}(b - a)^2 f''(c'),$$

with c' a point between a and b. Solving for $f'(a)$ above, we get

$$f'(a) = \frac{1}{(b - a)}\big(f(b) - f(a)\big) - \frac{1}{2}(b - a)f''(c'),$$

which replaced in the first equation gives

$$f(x) = f(a) + (x - a)\frac{1}{(b - a)}\big(f(b) - f(a)\big)$$
$$- \frac{1}{2}(x - a)(b - a)f''(c') + \frac{1}{2}(x - b)^2 f''(c).$$

Moreover, since $(x - a)(b - a)$ and $(x - b)^2$ are ≥ 0 on $[a, b]$, with $m'' = \inf_{[a,b]} f''$ and $M'' = \sup_{[a,b]} f''$, it readily follows that

$$-\frac{1}{2}(x - a)(b - a)M'' + \frac{1}{2}(x - b)^2 m''$$

$$\leq f(x) - f(a) + (x - a)\frac{1}{(b - a)}\big(f(b) - f(a)\big)$$

$$\leq -\frac{1}{2}(x - a)(b - a)m'' + \frac{1}{2}(x - b)^2 M''.$$

Also, since

$$\int_{[a,b]}(x - a)\,dx = \frac{(b - a)^2}{2} \quad \text{and} \quad \int_{[a,b]}(x - b)^2\,dx = \frac{(b - a)^3}{3},$$

integrating over $[a, b]$, we get

$$-\frac{1}{4}(b - a)^3 M'' + \frac{1}{6}(b - a)^3 m''$$

$$\leq \int_{[a,b]} f - (b - a)\frac{\big(f(a) + f(b)\big)}{2}$$

$$\leq -\frac{1}{4}(b - a)^3 m'' + \frac{1}{6}(b - a)^3 M''.$$

These are the bounds for the simple trapezoid rule. As for the compound trapezoid rule, let $\mathcal{P}_n = \{I_k^n\}$ be a partition of I in Π_e. Considering the above expression for $[a, b] = I_k^n$, $1 \leq k \leq n$, and summing over k and multiplying through by n^2, we have

$$-\frac{1}{4}U_{\Pi_e}(f, \mathcal{P}_n) + \frac{1}{6}L_{\Pi_e}(f, \mathcal{P}_n)$$

$$\leq n^2\left(\int_I f - \frac{1}{n}\sum_{k=1}^n \frac{\big(f(x_{k-1}^n) + f(x_k^n)\big)}{2}\right) = n^2\,\Delta_{T,n}(f)$$

$$\leq -\frac{1}{4}L_{\Pi_e}(f, \mathcal{P}_n) + \frac{1}{6}U_{\Pi_e}(f, \mathcal{P}_n),$$

which, since $\lim_n L_{\Pi_e}(f, \mathcal{P}_n) = \lim_n U_{\Pi_e}(f, \mathcal{P}_n) = \int_I f''$, implies that

$$\lim_n n^2 \Delta_{T,n}(f) = -\frac{1}{12}\int_I f'' = -\frac{1}{12}(f'(1) - f'(0)).$$

This completes the proof.   □

It is also possible to compute the error in the trapezoid rule exactly for functions that are expressed as a trigonometric series. In particular, for even functions defined on $[-\pi, \pi]$, it turns out to be convenient to use a Fourier cosine series on $[0, \pi]$,

$$f(x) = \frac{a_0}{2} + \sum_{k=1}^{\infty} a_k \cos(kx),$$

where the coefficients a_k are given by

$$a_k = \frac{2}{\pi} \int_{[0,\pi]} f(x) \cos(kx)\, dx, \quad k = 0, 1, 2, \ldots$$

For any $f(x)$ whose series converges to $f(x)$ at every x in $[0, \pi]$, by the Riemann–Lebesgue lemma, $a_k \to 0$ for $k \to \infty$.

Now, the integral $\int_{[0,\pi]} f = (\pi/2)\, a_0$. As for the expression $S_{T,n}(f)$, it is equal to

$$S_{T,n}(f) = \int_{[0,\pi]} f + \frac{\pi}{n} \sum_{k=1}^{\infty} a_k \left(\frac{1 + \cos(k\pi)}{2} + \sum_{N=1}^{n-1} \cos(Nk\pi/n) \right).$$

By the Lagrange expression (24), the only time the factor multiplying a_k is nonzero is when $k = 2mn$ where m is any integer (i.e., when k is an even multiple of n), in which case it is equal to n. Hence,

$$\Delta_{T,n}(f) = -\pi \sum_{m=1}^{\infty} a_{2mn}.$$

Thus, the convergence rate of the Fourier cosine series determines the approximation rate of the trapezoidal rule [50].

Now, an examination of Theorems 4 and 5 intuitively suggests that since the error in the midpoint rule is half that of the error in the trapezoidal rule and of opposite sign, the linear combination

$$SI(f, n) = \frac{1}{3}\big(T(f, n) + 2M(f, n)\big)$$

$$= \frac{1}{6} \frac{(b-a)}{n} \sum_{k=1}^{n} \left(f(x_{k-1}) + 4f\left(\frac{x_{k-1} + x_k}{2}\right) + f(x_k) \right)$$

corresponding to Simpson's rule, should be a better approximation.

A word about the self-taught mathematician Thomas Simpson (1710–1761). Born the son of a weaver, he was kicked out of the house by his father because his reading and studying interfered with his weaving. He moved in with a widow (the mother of one of his friends), and they were married in 1730. Through the books an astrologer fellow lodger lent him, Simpson learned the art and became a renowned fortune teller. A young woman approached him to find out about her lover who was at sea, and not to disappoint her, an assistant of Simpson appeared as a vision, and the woman suffered such a shock that she lost her mind for a time. Simpson quit his fortune telling, and moved to Derby, where he worked as a weaver and opened a school. In 1737, he published *A new Treatise of Fluxions*, a high-quality textbook devoted to the calculus of fluxions, republished in 1750 as *The doctrine and applications of Fluxions*. The topic was advanced for the time, and it was no trivial feat to write such a book when calculus was mastered by only a few mathematicians in Europe.

Simpson was also part of a group of itinerant lecturers who taught in London coffee houses. At this time, coffee houses were sometimes called the Penny Universities because of the cheap education they provided. They would charge an entrance fee of one penny, and customers would listen to lectures while having coffee. Different coffee houses catered to specific interests such as art, business, law, and mathematics.

Simpson is best remembered for his work on interpolation and numerical integration. Simpson's rule, although it did appear in his work, was something he learned from Newton's work as he himself acknowledged. Some have attributed the rule to J. Kepler (1571–1630), and in Germany, it is often referred to as *Keplersche Fassregel* (Kepler's barrel rule). Kepler's interest in calculating areas and volumes stemmed from the time in Linz, Austria, in 1613, when he had purchased a barrel of wine for his second wedding and the wine merchant's method of measuring the volume angered him. This inspired Kepler to study how to calculate areas and volumes and to write a book about the subject. Others attribute it to B.Cavalieri (1598–1647). On the other hand, the modern iterative form $x_{n+1} = x_n - f(x_n)/f'(x_n)$ in Newton's method of finding a root of a nonlinear function, derived below, is due to Simpson, who published it in 1740.

Returning to our discussion, let

$$\Delta_{SI,n}(f) = \int_I f - SI(f, n)$$

and recall that for $n = 1$,

$$SI(f, 1) = \frac{b - a}{6}\left(f(a) + 4f\left(\frac{a + b}{2}\right) + f(b)\right)$$

is the expression in (10), and, as noted there, the simple Simpson rule is exact up to polynomials of third degree or less. We might thus expect the error to be expressed in terms of fourth order derivatives, and this is indeed the case, as a familiar error expression in this case assumes the form $-\left((b - a)^5/2880\right) f^{(iv)}(\xi)$, [25].

However, there are simple functions, such as $f_p(x) = (x - a)^p$, with $p \in (3, 4)$, that satisfy $\lim_{x \to a^+} f_p^{(iv)}(x) = \infty$, and therefore, one of the most used quadrature formulae in practical applications does not apply to them. Yet, it is possible to estimate the error in terms of lower order derivatives [30, 76], as the following result shows:

Theorem 6 *Let f be a function defined on [a, b], and suppose that f has Riemann integrable derivatives up to order m there, where m = 1, 2, 3, 4. Then the following inequalities hold:*

$$\left| \Delta_{SI,1}(f) \right| \le C_m \, (b-a)^m \int_{[a,b]} |f^{(m)}|, \quad m = 1, 2, 3, 4,$$

where $C_1 = 1/3$, $C_2 = 1/24$, $C_3 = 1/324$, and $C_4 = 1/1152$.

Proof We will first consider the case $m = 1$. For an interval $I = [a, b]$, let $I_l = [a, (a+b)/2]$ and $I_r = [(a+b)/2, b]$, denote the left half and the right half of I, respectively.
  Let

$$s_1(x) = \left(x - \frac{5a+b}{6} \right) \chi_{I_l}(x) + \left(x - \frac{a+5b}{6} \right) \chi_{I_r}(x).$$

Then,

$$\int_{[a,b]} s_1 \, f' = \int_{[a,(a+b)/2]} \left(x - \frac{5a+b}{6} \right) f'(x)\,dx$$

$$+ \int_{[(a+b)/2,b]} \left(x - \frac{a+5b}{6} \right) f'(x)\,dx$$

$$= \left(x - \frac{5a+b}{6} \right) f(x) \Big]_a^{(a+b)/2}$$

$$+ \left(x - \frac{a+5b}{6} \right) f(x) \Big]_{(a+b)/2}^b - \int_{[a,b]} f.$$

Now, a straightforward computation gives that the integrated term is equal to $SI(f, 1)$, and, consequently, $\Delta_{SI,1}(f) = \int_I s_1 f'$. This expression is then bounded by

$$\int_{[a,b]} |s_1| \, |f'| \le \frac{1}{3} (b-a) \int_{[a,b]} |f'|.$$

The proof of the remaining cases follows along similar lines using the functions [38],

$$s_2(x) = \frac{1}{2}(x-a)\left(x - \frac{2a+b}{3} \right) \chi_{I_l}(x) + \frac{1}{2}(x-b)\left(x - \frac{a+2b}{3} \right) \chi_{I_r}(x),$$

$$s_3(x) = -\frac{1}{6}(x-a)^2\left(x - \frac{a+b}{2}\right)\chi_{I_l}(x) - \frac{1}{6}(x-b)^2\left(x - \frac{a+b}{2}\right)\chi_{I_r}(x),$$

and

$$s_4(x) = \frac{1}{24}(x-a)^3\left(x - \frac{a+2b}{3}\right)\chi_{I_l}(x) + \frac{1}{24}(x-b)^3\left(x - \frac{2a+b}{3}\right)\chi_{I_r}(x),$$

and the relations

$$\Delta_{SI,1}(f) = \int_I s_2 f'' = \int_I s_3 f''' = \int_I s_4 f^{(iv)}.$$

This completes the proof.   □

There is a trick left in our bag. Since $\int_I s_1 = 0$, we can improve the estimate of $\int_I s_1 f'$ by subtracting a constant from f'. In this case, we have:

Theorem 7 *Let f be a function defined on $I = [a,b]$, with a derivative f' that is Riemann integrable there. Then, with $m_I = \inf_I f'(x)$ and $M_I = \sup f'(x)$, it follows that*

$$\left|\Delta_{SI,1}(f)\right| \le \frac{1}{6}(M_I - m_I)(b-a)^2.$$

Moreover, if f'^2 is Riemann integrable on I,

$$\left|\Delta_{SI,1}(f)\right| \le \frac{1}{6}(b-a)^{3/2}\left(\int_I f'^2 - \frac{\left(f(b) - f(a)\right)^2}{b-a}\right)^{1/2}. \tag{21}$$

Inequality (21) is sharp in the sense that $1/6$ cannot be replaced by a smaller constant there.

Furthermore, if $\mathcal{P}_n = \{I_k^n\}$, $I_k^n = [x_{k,l}, x_{k,r}]$, $1 \le k \le n$, is a partition of $[a,b]$ in Π_e,

$$\left|\Delta_{SI,n}(f)\right| \le \frac{1}{6}\frac{b-a}{n}\sum_{k=1}^{n}\left(\frac{b-a}{n}\int_{I_k^n} f'^2 - \left(f(x_{k,r}) - f(x_{k,l})\right)^2\right)^{1/2}.$$

Proof As noted in Theorem 6 we have $\Delta_{SI,1}(f) = \int_I s_1 f'$. Then, since $\int_I s_1 = 0$, $\int_I s_1 f' = \int_I s_1(f' - c)$ for an arbitrary constant c, and since we have $\sup_{x\in I}|s_1(x)| = (b-a)/3$, it follows that

$$\left|\int_I s_1 f'\right| \le \frac{(b-a)}{3}\int_I |f' - c|.$$

Now, $f'(x) - m_I \geq 0$ throughout I, and so with $S_I = (f(b) - f(a))/(b - a)$, we have

$$\int_I |f' - m_I| = \int_I (f' - m_I) = (b - a)S_I - m_I(b - a),$$

and, consequently,

$$\left| \int_I s_1 f' \right| \leq \frac{1}{3}(S_I - m_I)(b - a)^2.$$

Also, since $f'(x) \leq M_I$ throughout I, a similar computation gives

$$\left| \int_I s_1 f' \right| \leq \frac{1}{3}(M_I - S_I)(b - a)^2,$$

and adding these estimates gives

$$\left| \int_I s_1 f' \right| \leq \frac{1}{6}(M_I - m_I)(b - a)^2.$$

Let next f'_I denote the average of f' over the interval I. Since for a constant c we have

$$|f'(x) - f_I| \leq |f'(x) - c| + |c - f'_I| \leq |f'(x) - c| + \frac{1}{|I|} \int_I |f'(y) - c| \, dy,$$

integrating over I, it follows that $\int_I |f'(x) - f'_I| \, dx \leq 2 \int_I |f'(x) - c| \, dx$, and, consequently,

$$\inf_c \int_I |f'(x) - c| \, dx \leq \int_I |f'(x) - f'_I| \, dx \leq 2 \inf_c \int_I |f'(x) - c| \, dx.$$

Thus we can do no better than subtracting f'_I. Then, by the Cauchy–Schwartz inequality,

$$\left| \int_{[a,b]} s_1 f' \right| = \left| \int_{[a,b]} s_1 (f' - f'_I) \right| \leq \left(\int_I s_1^2 \right)^{1/2} \left(\int_I (f' - f'_I)^2 \right)^{1/2},$$

which in turn, since

$$\int_I s_1^2 = \frac{(b - a)^3}{36},$$

and

$$\int_I (f' - f_I')^2 = \int_I f'^2 - \frac{(f(b) - f(a))^2}{b - a},$$

yields (21).

To verify the sharpness of the inequality, consider $I = [0, 1]$, and the function

$$f(x) = \left(\frac{1}{2}x^2 - \frac{1}{6}x\right) \chi_{[0,1/2)}(x) + \left(\frac{1}{2}x^2 - \frac{5}{6}x + \frac{1}{3}\right) \chi_{[1/2,1]}(x).$$

Then, $\int_I f = 0$, $f(0) = f(1) = 0$, $f(1/2) = 1/24$, and $\int_I f'^2 = 1/36$, so that if (21) holds with C in place of $1/6$, there we have

$$\frac{1}{36} = \left|\frac{1}{3} 2f\left(\frac{1}{2}\right)\right| \le C \left(\int_I f'^2\right)^{1/2} = \frac{1}{6} C.$$

Hence, $6C \ge 1$, and $1/6$ is the best possible constant in (21), [103].

Finally, let $\mathcal{P}_n = \{I_k^n\}$, $I_k^n = [x_{k,l}, x_{k,r}]$, $1 \le k \le n$, be a partition of $[a, b]$ in Π_e. Then by (21), since $x_{k,r} - x_{k,l} = (b - a)/n$ for all k,

$$\left|\int_{I_k^n} f - \frac{1}{6} \frac{b - a}{n}\left(f(x_{k,l}) + 4f\left(\frac{x_{k,l} + x_{k,r}}{2}\right) + f(x_{k,r})\right)\right|$$

$$\le \frac{1}{6} \frac{(b - a)^{3/2}}{n^{3/2}}\left(\int_{I_k^n} f'^2 - n \frac{(f(x_{k,r}) - f(x_{k,l}))^2}{b - a}\right)^{1/2}.$$

Then, summing and simplifying, it readily follows that

$$|\Delta_{SI,n}(f)| \le \frac{1}{6} \frac{b - a}{n} \sum_{k=1}^{n} \left(\frac{b - a}{n} \int_{I_k^n} f'^2 - (f(x_{k,r}) - f(x_{k,l}))^2\right)^{1/2},$$

and the proof is complete.   □

The situation is more manageable for smoother f, e.g., when the derivatives of f are continuous rather than merely Riemann integrable on I. For the trapezoid rule, by the mean value theorem for integrals, from Proposition 7, it follows that for some $\xi \in [a, b]$, the error expression is

$$\Delta_{T,1}(f) = -\frac{1}{2} \int_I (b - x)(x - a) f''(x) \, dx = -\frac{1}{12}(b - a)^3 f''(\xi).$$

We have stated above a similar error expression for the Simpson rule. And, for the left Riemann sums, since

$$\int_{[a,b]} f = -\int_{[a,b]} f(t)d(b-t) = -f(t)(b-t)]_a^b + \int_{[a,b]} f'(t)(b-t)\,dt$$

$$= f(a)(b-a) + \frac{(b-a)^2}{2} f'(\xi),$$

it readily follows that there is $\xi \in I$ such that

$$\Delta_{L,1}(f) = \frac{(b-a)^2}{2} f'(\xi).$$

In general, an integration rule whose error expression $E(f)$ on $I = [a,b]$ assumes the form

$$E(f) = M (b-a)^{k+1} f^{(k)}(\xi), \quad \xi \in I,$$

whenever $f \in C^k(I)$ is said to be a *simplex* on I, [25]. The constant M may depend on the rule but is otherwise independent of I and f. In fact, all the elementary numerical quadrature methods satisfy this condition [34, 93].

We claim that if we now consider the compound rule that arises when I is divided into n equal length subintervals and let $E_n(f)$ designate its error, we have for $f \in C^k(I)$ the asymptotic relation

$$\lim_n n^k E_n(f) = M(b-a)^k (f^{(k-1)}(b) - f^{(k-1)}(a)).$$

To see this, let $[x_{j-1}, x_j]$ denote the jth subinterval of I and $E_{j,R}(f)$ the corresponding error, and observe that by linearity, transferring the error rule to these subintervals, it follows that there are ξ_j with $x_{j-1} \le \xi_j \le x_j$ for $1 \le j \le n$, such that

$$E_n(f) = \sum_{j=1}^n E_{j,R}(f) = \sum_{j=l}^n M \frac{(b-a)^{k+1}}{n^{k+1}} f^{(k)}(\xi_j)$$

$$= M \frac{(b-a)^k}{n^k} \frac{(b-a)}{n} \sum_{j=1}^n f^{(k)}(\xi_j).$$

Now, since

$$\lim_n \frac{(b-a)}{n} \sum_{j=1}^n f^{(k)}(\xi_j) = \int_{[a,b]} f^{(k)} = (f^{(k-1)}(b) - f^{(k-1)}(a)),$$

the conclusion follows.

For instance, in the specific case of the trapezoid rule, we get

$$\lim_n n^2 \Delta_{T,n}(f) = -\frac{1}{12}(b-a)^2 (f'(b) - f'(a)),$$

which is precisely the conclusion of Theorem 5. And for the left Riemann sums, we get

$$\lim_n n \Delta_{L,n}(f) = \frac{(b-a)}{2} (f(b) - f(a)),$$

which is the conclusion of Theorem 2.

2.4   Roots of Nonlinear Equations

We discuss next an application of quadrature methods to the calculation of the roots of a nonlinear equation. Newton's iterative approach, perhaps the best known and most widely used algorithm, can be derived, and improved, using quadrature methods [107].

The setting is that of a smooth function f defined on an open interval D, and assumed to have a simple zero at a point $a \in D$, i.e., $f(a) = 0$ and $f'(a) \neq 0$. The simple left Riemann sum of f' is given by

$$\int_{[x_n,x]} f' \approx (x - x_n) f'(x_n), \quad x_n, x \in D,$$

and since $\int_{[x_n,x]} f' = f(x) - f(x_n)$, we have

$$f(x) \approx f(x_n) + (x - x_n) f'(x_n).$$

Now, if a is a zero of f, it follows that

$$a \approx x_n - \frac{f(x_n)}{f'(x_n)},$$

which suggests the value of the Newton iterates x_{n+1} as

$$x_{n+1} = x_n - \frac{f(x_n)}{f'(x_n)}, \quad n = 0, 1, \ldots,$$

where x_0 is assumed to be chosen sufficiently close to a, and which converges to the root quadratically. Informally, this means that after some iterations, the process doubles the number of correct decimal places or significant digits at each iteration.

Weerakoon and Fernando [107] proposed a variant of Newton's method by applying instead the simple trapezoidal rule to f',

$$\int_{[x_n,x]} f' \approx \frac{(x-x_n)}{2}\left(f'(x_n)+f'(x)\right).$$

This yields

$$f(x) \approx f(x_n) + \frac{(x-x_n)}{2}\left(f'(x_n)+f'(x)\right),$$

which at a zero a of f becomes

$$a \approx x_n - \frac{2f(x_n)}{f'(x_n)+f'(a)}$$

and suggests the new iterate value x_{n+1}, this time given implicitly, as

$$x_{n+1} = x_n - \frac{2f(x_n)}{f'(x_n)+f'(x_{n+1})}.$$

In order to deal with the implicitness, we replace x_{n+1} on the right-hand side above with the value of the Newton iterate and get

$$x_{n+1} = x_n - \frac{2f(x_n)}{f'(x_n)+f'\left(x_n - \frac{f(x_n)}{f'(x_n)}\right)}, \quad n = 0, 1, 2, \ldots$$

as the value for the successive iterates, where x_0 is assumed to have been chosen sufficiently close to the root.

In order to carry out the analysis of the convergence of $\{x_n\}$, we will make use of Taylor's formula for f in a neighbourhood of the zero a in D of f. We adopt the usual notation in this context, namely,

$$e_n = x_n - a, \quad \text{and,} \quad C_k = \frac{1}{f'(a)}\frac{1}{k!}f^{(k)}(a), \quad k = 1, 2, \ldots$$

We then have:

Proposition 8 *Let $f : D \to R$ be defined on an open interval D, and assume that f has first, second, and third derivatives in the interval D. If f has a simple root at $a \in D$ and x_0 is sufficiently close to a, then the iterations for the variant Newton method satisfy the error equation,*

$$e_{n+1} = \left(C_2^2 + \frac{1}{2}C_3\right)e_n^3 + O(e_n^4).$$

Proof Since a is a simple root of f, the Taylor expansion at x_n reads

$$f(x_n) = f(a + e_n) = f'(a)\Big(e_n + C_2\,e_n^2 + +C_3\,e_n^3 + O(e_n^4)\Big),$$

and that of f' at x_n reads

$$f'(x_n) = f'(a)\Big(1 + 2\,C_2\,e_n + 3\,C_3\,e_n^2 + O(e_n^3)\Big),$$

and, therefore, long division gives

$$\frac{f(x_n)}{f'(x_n)} = e_n - C_2\,e_n^2 + (2\,C_2 - 2\,C_3)\,e_n^3 + O(e_n^4).$$

Whence the Newton iterates x_{n+1}^* verify,

$$x_{n+1}^* = x_n - \frac{f(x_n)}{f'(x_n)} = a + C_2\,e_n^2 + (2\,C_3 - 2\,C_2^2)\,e_n^3 + O(e_n^4),$$

and, again by a Taylor's expansion at the Newton iterate x_{n+1}^*, it follows that

$$f'(x_{n+1}^*) = f'(a)\Big(1 + 2\,C_2^2\,e_n^2 + 4\,C_2(C_3 - C_2^2)\,e_n^3 + O(e_n^4)\Big),$$

which added to the expansion for $f'(x_n)$ gives

$$f'(x_n) + f'(x_{n+1}) = 2\,f'(a)\Big(1 + C_2\,e_n + \Big(C_2^2 + \frac{3}{2}\,C_3\Big)e_n^2 + O(e_n^3)\Big).$$

Therefore,

$$\frac{2f(x_n)}{f'(x_n) + f'(x_{n+1}^*)} = e_n - \Big(C_2^2 + \frac{1}{2}\,C_3\Big)e_n^3 + O(e_n^4).$$

Thus, since for the successive iterates, we have

$$x_{n+1} = x_n - \frac{2f(x_n)}{f'(x_n) + f'(x_{n+1}^*)},$$

and $x_{n+1} = e_{n+1} + a$ and $x_n = e_n + a$, it readily follows that

$$e_{n+1} + a = e_n + a - (e_n - (C_2^2 + \frac{1}{2}C_3)e_n^3 + O(e_n^4))$$

and, consequently,

$$e_{n+1} = (C_2^2 + \frac{1}{2}C_3)e_n^3 + O(e_n^4),$$

thus establishing the third-order convergence of the variant Newton method considered. □

A feature of the variant Newton method is that unlike other third or higher order methods, no computations of second or higher derivatives are required to carry out the iterations. And it is economical too. For the function $f(x) = x^2 \sin^2(x) + \exp(x^2 \cos(x) \sin(x)) - 28$, starting out with $x_0 = 5$ and aiming at an approximation to the root to 15 decimal places (the desired value is then 4.82458931731526), Newton's method requires 9 iterations and 18 function evaluations to reach a computational order of convergence of 1.99 (close to quadratic), whereas the variant requires 5 iterations and 15 function evaluations to reach a computational order of convergence 2.87 (close to cubic) [107].

Frontini and Sormani showed that a similar result holds, independently from the quadrature formula used for the computation of the integral above [36].

Monotonicity of Riemann Sums

We will now address the question of the monotonicity of the Riemann sums. For functions f defined on $I = [0, 1]$, Bennett and Jameson considered conditions under which various averages of the values of f evaluated at n equally spaced points through I increase or decrease with n, [8]. In particular, they discussed the averages excluding and including the endpoints of I given, respectively, by

$$A_n(f) = \frac{1}{n-1} \sum_{k=1}^{n-1} f\left(\frac{k}{n}\right), \quad n \geq 2, \tag{22}$$

and

$$B_n(f) = \frac{1}{n+1} = \sum_{k=0}^{n} f\left(\frac{k}{n}\right), \quad n \geq 0. \tag{23}$$

One endpoint averages, such as the left and right Riemann sums of f, are also of interest. We begin with an observation of general nature:

Proposition 9 *Let f be a monotone function defined on I with $f(0) \neq f(1)$, and let $\mathcal{P}_n$ denote the partitions of I in Π_e. Then, if $S_R(f, \mathcal{P}_n) \leq S_R(f, \mathcal{P}_{n+1})$ for some n, it follows that $S_L(f, \mathcal{P}_n) < S_L(f, \mathcal{P}_{n+1})$ for the same n. On the other hand, if $S_R(f, \mathcal{P}_{n+1}) \leq S_R(f, \mathcal{P}_n)$ for some n, then, for the same n, $S_L(f, \mathcal{P}_{n+1}) < S_L(f, \mathcal{P}_n)$.*

Proof Assume first that f is increasing. Writing the assumption $S_R(f, \mathcal{P}_n) \leq S_R(f, \mathcal{P}_{n+1})$ in full, it readily follows that

$$f(1) \leq n \sum_{k=1}^{n} f(k/(n+1)) - (n+1) \sum_{k=1}^{n-1} f(k/n). \qquad (24)$$

For the sake of argument, suppose that $S_L(f, \mathcal{P}_{n+1}) \leq S_L(f, \mathcal{P}_n)$. A similar computation gives that

$$n \sum_{k=1}^{n} f(k/(n+1)) - (n+1) \sum_{k=1}^{n-1} f(k/n) \leq f(0),$$

which combined with (24) implies that $f(1) < f(0)$, which is not the case since f increases.

Now, if f is decreasing, $-f$ is increasing, $S_R(-f, \mathcal{P}_n) = -S_R(f, \mathcal{P}_n)$ and $S_L(-f, \mathcal{P}_n) = -S_L(f, \mathcal{P}_n)$. Hence, if $S_R(f, \mathcal{P}_{n+1}) \leq S_R(f, \mathcal{P}_n)$, from what we just proved it readily follows that $S_L(f, \mathcal{P}_{n+1}) \leq S_L(f, \mathcal{P}_n)$.

Moreover, with f still decreasing, if $S_R(f, \mathcal{P}_n) \leq S_R(f, \mathcal{P}_{n+1})$, the reverse inequality to (24) holds, and

$$n \sum_{k=1}^{n} f(k/(n+1)) - (n+1) \sum_{k=1}^{n-1} f(k/n) \leq f(1).$$

And, were $S_L(f, \mathcal{P}_{n+1}) \leq S_L(f, \mathcal{P}_n)$, a similar computation combined with the above remark gives that $f(0) < f(1)$, which is not the case since f is decreasing. To complete the proof, we just replace f by $-f$ in the last conclusion. □

Although numerical computations may suggest that $\{S_R(f, \mathcal{P}_n)\}$ and $\{S_L(f, \mathcal{P}_n)\}$ are monotone, it is not immediate to find applicable conditions that assure this. For instance, for the function $f(x) = x$, we have that $U(f, \mathcal{P}_n) = S_R(f, \mathcal{P}_n) = 1/2 + 1/2n$ is decreasing and that $L(f, \mathcal{P}_n) = S_L(f, \mathcal{P}_n)) = 1/2 - 1/2n$ is increasing, but this does not follow from any general principle. As it turns out, the monotone behaviour is an exception rather than a general phenomenon, and the convexity or concavity of f is present in all the cases criteria can be stated [58].

Recall that a function f defined on I is said to be *convex* on I if

$$f(\alpha x + (1-\alpha) y) \leq \alpha f(x) + (1-\alpha) f(y) \qquad (25)$$

for all $x, y \in I$ and $0 \leq \alpha \leq 1$. And, a function f is said to be *concave* on I if

$$f(\alpha x + (1-\alpha) y) \geq \alpha f(x) + (1-\alpha) f(y)$$

for all $x, y \in I$ and $0 \leq \alpha \leq 1$. Thus, f is convex iff $-f$ is concave.

We begin by considering the averages excluding the endpoints:

Proposition 10 *Let $A_n(f)$ be given by (22). If f is convex on $(0, 1)$, the sequence $\{A_n(f)\}$ is increasing. And, if f is concave on $(0, 1)$, $\{A_n(f)\}$ is decreasing.*

Proof Suppose first that f is convex on $(0, 1)$. Let then $x_h^m = h/m$, $1 \le h \le m$, $m = 1, 2, \ldots$ Now, for $1 \le k \le n$, x_k^n lies between x_k^{n+1} and x_{k+1}^{n+1}, and, consequently,

$$x_k^n = \alpha_k^n x_k^{n+1} + (1 - \alpha_k^n) x_{k+1}^{n+1}, \quad 1 \le k \le n, \tag{26}$$

where $0 < \alpha_k^n = (1 - k/n) < 1$. Then, from (26) and the convexity of f, it follows that

$$f(x_k^n) \le \left(1 - \frac{k}{n}\right) f(x_k^{n+1}) + \frac{k}{n} f(x_{k+1}^{n+1}),$$

and, consequently, summing,

$$\sum_{k=1}^{n-1} f(x_k^n) \le \left[\left(1 - \frac{1}{n}\right) f(x_1^{n+1}) + \frac{1}{n} f(x_2^{n+1})\right]$$

$$+ \left[\left(1 - \frac{2}{n}\right) f(x_2^{n+1}) + \frac{2}{n} f(x_3^{n+1})\right] + \cdots$$

$$+ \left[\left(1 - \frac{(n-1)}{n}\right) f(x_{n-1}^{n+1}) + \frac{n-1}{n} f(x_n^{n+1})\right]$$

$$= \frac{n-1}{n} \sum_{k=1}^{n} f(x_k^{n+1}) = (n-1) A_{n+1}(f),$$

which is the desired conclusion in this case.

And, if f is concave on $(0, 1)$, $-f$ is convex there, and by the result we just proved, $-A_n(f) = A_n(-f) \le A_{n+1}(-f) = -A_{n+1}(f)$. Hence, $\{A_n(f)\}$ decreases with n, and we have finished. $\square$

The corresponding result for the $B_n(f)$ is the following:

Proposition 11 *Let $B_n(f)$ be given by (23). If f is convex on $(0, 1)$, the sequence $\{B_n(f)\}$ is decreasing. And, if f is concave on $(0, 1)$, $\{B_n(f)\}$ is increasing.*

Proof It suffices to prove the result when f is convex. Let $x_h^m = h/m$, $1 \le h \le m$, $m = 1, 2, \ldots$ For $0 < k < n$, the point x_k^n lies between x_{k-1}^{n-1} and x_k^{n-1}, and for $0 < \alpha_k^n = k/n < 1$, $1 \le k \le n - 1$, we have

$$x_k^n = \alpha_k^n x_{k-1}^{n-1} + (1 - \alpha_k^n) x_k^{n-1} = \left(\frac{k}{n}\right) x_{k-1}^{n-1} + \left(1 - \frac{k}{n}\right) x_k^{n-1}. \tag{27}$$

Consistent with (27), note that $x_0^n = x_0^{n-1}$ and $x_n^n = x_{n-1}^{n-1}$, and for the values $k = 0, n$, we have $f(x_0^n) = (n/n) f(x_0^{n-1})$ and $f(x_n^n) = (n/n) f(x_{n-1}^{n-1})$, respectively. For the remaining values of k by the convexity of f from (27), it follows that

$$f(x_k^n) \leq \left(\frac{k}{n}\right) f(x_{k-1}^{n-1}) + \frac{n-k}{n} f(x_k^{n-1}).$$

Hence, combining these assertions,

$$\sum_{k=0}^n f(x_k^n) \leq \frac{n}{n} f(x_0^{n-1}) + \left[\frac{1}{n} f(x_0^{n-1}) + \frac{n-1}{n} f(x_1^{n-1})\right]$$

$$+ \left[\frac{2}{n} f(x_1^{n-1}) + \frac{n-2}{n} f(x_2^{n-1})\right] + \cdots$$

$$+ \left[\frac{n-2}{n} f(x_{n-2}^{n-1}) + \frac{1}{n} f(x_{n-1}^{n-1})\right] + \frac{n}{n} f(x_{n-1}^{n-1})$$

$$= \frac{n+1}{n} \sum_{k=0}^{n-1} f(x_k^{n-1}) = (n+1) B_{n-1}(f),$$

which implies that $B_n(f) \leq B_{n-1}(f)$, and the proof is finished.   □

For a monotone function f that is either convex or concave on I, we are interested in the monotonicity of the right and left Riemann sums of f. Taking into account the monotonicity and convexity of f, there are eight possible outcomes for the monotonicity of these Riemann sums of f, and they are covered in our next result. A useful observation concerning these sums is

$$S_R(f(1 - \cdot), \mathcal{P}_n) = S_L(f, \mathcal{P}_n), \quad \text{and,} \quad S_L(f(1 - \cdot), \mathcal{P}_n) = S_R(f, \mathcal{P}_n). \quad (28)$$

We then have:

Theorem 8 *Let f be a monotone function defined on I that is either concave or convex on $(0, 1)$. Then, if f is increasing on I, $\{S_L(f, \mathcal{P}_n)\}$ increases and $\{S_R(f, \mathcal{P}_n)\}$ decreases with n. And, if f is decreasing on I, then $\{S_R(f, \mathcal{P}_n)\}$ increases and $\{S_L(f, \mathcal{P}_n)\}$ decreases with n.*

Proof Suppose first that f is convex increasing on $(0, 1)$, and let $g(x) = f(x) - f(0)$. Then g is positive, increasing in I, and convex on $(0, 1)$. Also, $g(0) = 0$ and $A_n(g) \geq 0$ for all n. Then,

$$S_L(g, \mathcal{P}_n) = \frac{1}{n} \sum_{k=0}^{n-1} g(k/n) = \frac{1}{n} \sum_{k=1}^{n-1} g(k/n) = \left(1 - \frac{1}{n}\right) A_n(g). \quad (29)$$

Now, since $(1 - 1/n) \leq (1 - 1/(n + 1))$ and by Proposition 10, the sequence $\{A_n(g)\}$ increases with n, the right-hand side of (29) does not exceed

$$\left(1 - \frac{1}{n + 1}\right) A_{n+1}(g) = S_L(g, \mathcal{P}_{n+1}),$$

and, consequently, $\{S_L(g, \mathcal{P}_n)\}$ increases with n.

Moreover, since $S_L(g, \mathcal{P}_n) = S_L(f, \mathcal{P}_n) - f(0)$, it follows that

$$S_L(f, \mathcal{P}_n) - f(0) = S_L(g, \mathcal{P}_n) \leq S_L(g, \mathcal{P}_{n+1}) = S_L(f, \mathcal{P}_{n+1}) - f(0),$$

and, consequently, $S_L(f, \mathcal{P}_n) \leq S_L(f, \mathcal{P}_{n+1})$, all $n \geq 2$ (strictly, unless f is constant.) Since as is readily verified $S_L(f, \mathcal{P}_1) \leq S_L(f, \mathcal{P}_2)$, the proof is complete in this case.

Now, if f is convex decreasing on $(0, 1)$, $f(1 - x)$ is convex increasing on $(0, 1)$ and by (28), $S_R(f, \mathcal{P}_n) = S_L(f(1 - \cdot), \mathcal{P}_n) \leq S_L(f(1 - \cdot), \mathcal{P}_{n+1}) = S_R(f, \mathcal{P}_{n+1})$, and so $\{S_R(f, \mathcal{P}_n)\}$ increases with n.

And, if f is concave decreasing on $(0, 1)$, then $-f$ is convex increasing on $(0, 1)$, $-S_L(f, \mathcal{P}_n) = S_L(-f, \mathcal{P}_n) \leq S_L(-f, \mathcal{P}_{n+1}) = -S_L(f, \mathcal{P}_{n+1})$, and $\{S_L(f, \mathcal{P}_n)\}$ decreases with n.

Finally, if f is concave increasing on $(0, 1)$, $-f$ is convex decreasing on $(0, 1)$, and, consequently, $-S_R(f, \mathcal{P}_n) = S_R(-f, \mathcal{P}_n) \leq S_R(-f, \mathcal{P}_{n+1}) = -S_R(f, \mathcal{P}_{n+1})$, and $\{S_R(f, \mathcal{P}_n)\}$ decreases with n.

To complete the proof of the remaining four statements, with f increasing and convex on $(0, 1)$ and g as above, observe that since $g(0) = 0$,

$$S_R(g, \mathcal{P}_n) - S_L(g, \mathcal{P}, n) = \frac{g(1)}{n} = \left(1 + \frac{1}{n}\right) B_n(g) - \left(1 - \frac{1}{n}\right) A_n(g),$$

and, consequently, from (29), it follows that

$$S_R(g, \mathcal{P}_n) = \left(1 + \frac{1}{n}\right) B_n(g).$$

Now, since $(1 + 1/n) \geq (1 + 1/(n + 1))$ and by Proposition 11, the sequence $\{B_n(g)\}$ decreases with n, and since $B_n(g) \geq 0$, the right-hand side above dominates

$$\left(1 + \frac{1}{(n + 1)}\right) B_{n+1}(g) = S_R(g, \mathcal{P}_{n+1}),$$

which implies that $\{S_R(g, \mathcal{P}_n)\}$ decreases with n, and the same is true for $\{S_R(f, \mathcal{P}_n)\}$.

It then readily follows that if f is convex decreasing on $(0, 1)$, $\{S_L(f, \mathcal{P}_n)\}$ decreases. And, if f is concave decreasing on $(0, 1)$, that $\{S_R(f, \mathcal{P}_n)\}$ increases with n, and, lastly, if f is concave increasing on $(0, 1)$, $\{S_L(f, \mathcal{P}_n)\}$ increases with n. Thus all eight possibilities have been covered, and the proof is complete. $\square$

We will now establish that Theorem 8 holds if the monotonicity assumption on f is dropped. We have:

Theorem 9 *Let f be a function defined on I. If f is convex on $(0, 1)$, $\{S_L(f, \mathcal{P}_n)\}$ increases with n.*

Proof If f is a convex function on $[0, 1]$ that is not monotone, f attains its minimum on $[0, 1]$ at some point $c \in (0, 1)$. Then, f is decreasing on $[0, c]$ and increasing on $[c, 1]$. Let $g(x) = f(x) - f(c)$, $g(c) = 0$. For partitions $\mathcal{P}_n = \{I_k^n\}$ of I in Π_e, let k_c denote the index k such that $c \in I_k$. Then the minimum of g on I_k is $g(k/n)$ for $k \le k_c - 1$, 0 for $k = k_c$, and $g((k-1)/n)$ for $k \ge k_c + 1$. Hence,

$$S_L(g, \mathcal{P}_n) = \left(1 - \frac{1}{n}\right) A_n(g),$$

which is the relation (29), and the proof proceeds as in Theorem 6.   □

The corresponding statement for $\{S_R(f, \mathcal{P}_n)\}$ is also true, but the proof is more elaborate [8].

In our next result, the tags are allowed to vary from the endpoints, while nevertheless the monotonicity of the Riemann sums is preserved. In this case, we have:

Theorem 10 *Let f be a function defined in a neighbourhood of $[0, 1]$, and with a real sequence $r = \{r_k^n\}$, put*

$$S_R^r(f, n) = \frac{1}{n} \sum_{k=1}^n f\left(\frac{k}{n} + r_k^n\right),$$

and, with $\alpha_k^n = (1 - k/n)$ as in (26), let

$$\beta_{k,r}^n = r_k^n - \alpha_k^n \, r_k^{n+1} - (1 - \alpha_k^n) \, r_{k+1}^{n+1}. \tag{30}$$

Suppose that $\beta_{k,r}^n \ge 0$ for all k, n. Then, if f is decreasing and convex on $(0, 1)$, $\{S_R^r(f, n)\}$ is increasing with n. And, if f is increasing and concave on $(0, 1)$, $\{S_R^r(f, n)\}$ is decreasing with n.

Proof Let $x_h^m = h/m$ and $x_{h,r}^m = h/m + r_h^m$, for $1 \le h \le m$, $m = 1, 2, \ldots$ Now, $x_k^n = \alpha_k^n \, x_k^{n+1} + (1 - \alpha_k^n) \, x_{k+1}^{n+1}$, and we will look for a similar relationship for the $x_{k,r}^n$. Indeed, we have

$$x_{k,r}^n = \alpha_k^n \, x_{k,r}^{n+1} + (1 - \alpha_k^n) \, x_{k+1,r}^{n+1} + \beta_{k,r}^n, \tag{31}$$

where $\beta_{k,r}^n$ is given by (30).

Since $\beta_k^n \geq 0$ for all n, for f decreasing and convex, from (31), it follows that for $k = 1, \ldots, n$,

$$f(x_{k,r}^n) \leq f\left(\left(1 - \frac{k}{n}\right)x_{k,r}^{n+1} + \frac{k}{n}x_{k+1,r}^{n+1}\right)$$

$$\leq \left(1 - \frac{k}{n}\right)f(x_{k,r}^{n+1}) + \frac{k}{n}f(x_{k+1,r}^{n+1}). \qquad (32)$$

Moreover, since f is decreasing, it readily follows that

$$\left(1 - \frac{k}{n}\right)f(x_{k,r}^{n+1}) + \frac{k}{n}f(x_{k+1,r}^{n+1})$$

$$\leq \left(1 - \frac{k}{(n+1)}\right)f(x_{k,r}^{n+1}) + \frac{k}{(n+1)}f(x_{k+1,r}^{n+1}). \qquad (33)$$

Thus, combining and adding (32) and (33), it follows that

$$(n+1)\sum_{k=1}^{n} f(x_{k,r}^n) \leq \sum_{k=1}^{n}(n+1-k)f(x_{k,r}^{n+1}) + \sum_{k=1}^{n} k\, f(x_{k+1,r}^{n+1}). \qquad (34)$$

Now, since $\sum_{k=1}^{n} k\, f(x_{k+1,r}^{n+1}) = \sum_{k=2}^{n+1}(k-1)f(x_{k,r}^{n+1})$, the sum on the right-hand side of (34) becomes

$$n\,f(x_{1,r}^{n+1}) + n\sum_{k=2}^{n} f(x_{k,r}^{n+1}) + nf(x_{n+1,r}^{n+1}) = n\sum_{k=1}^{n+1} f(x_{k,r}^{n+1}).$$

Hence, from (34), it follows that

$$(n+1)\sum_{k=1}^{n} f(x_{k,r}^n) \leq n\sum_{k=1}^{n+1} f(x_{k,r}^{n+1}),$$

which is what we wanted to show.

Now, if f is increasing and concave, $-f$ is convex decreasing, applying the above result, we get $-S_R^r(f,n) = S_R^r(-f,n) \leq S_R^r(-f,n+1) = -S_R^r(f,n+1)$, and so $\{S_R^r(f,n)\}$ decreases with n. Thus the proof is finished. □

When the double sequence $\{r_k^n\}$ depends only on the level n, i.e., $r_k^n = r_n$ for $1 \leq k \leq n$, Then $\beta_k^n = \beta_n = r_n - r_{n+1}$, which is positive when $\{r_n\}$ is decreasing. In particular, it holds for $r_n = r/n$, all n, [2].

We also have:

Theorem 11 *Let f be a function defined in a neighbourhood of* [0, 1], *and with a real sequence* $r = \{r_k^n\}$, *put*

$$S_L^r(f, n) = \frac{1}{n} \sum_{k=0}^{n-1} f\left(\frac{k}{n} + r_k^r\right), \quad n \geq 2,$$

and, with $\alpha_k^n = k/n$, *let*

$$\beta_k^n = r_k^n - \alpha_k^n r_{k-1}^{n-1} - (1 - \alpha_k^n) r_k^{n-1}. \tag{35}$$

Suppose that $\beta_k^n \leq 0$ *for all* k, n. *Then, if* f *is decreasing and convex on* (0, 1), $\{S_L^r(f, n)\}$ *is decreasing. And if* f *is concave and increasing, the sequence* $\{S_L^r(f, n)\}$ *is increasing.*

Proof Let $x_h^m = h/m$ and $x_{h,r}^m = h/m + r_h^m$, $1 \leq h \leq m$, $m = 1, 2, \ldots$, and note that with α_k^n as in (26),

$$x_{k,r}^n = \alpha_k^n x_{k-1,r}^{n-1} + (1 - \alpha_k^n) x_{k,r}^{n-1} + \beta_k^n, \tag{36}$$

where β_k^n is as in (35).

Suppose first that f is decreasing and convex. Since $\beta_k^n \leq 0$ for all k, n, from (36), it follows that

$$f(x_{k_r}^n) \leq \left(\frac{k}{n}\right) f(x_{k-1,r}^{n-1}) + \left(1 - \frac{k}{n}\right) f(x_{k,r}^{n-1}), \quad 1 \leq k \leq n - 1.$$

Now, since f is decreasing, the right-hand side of this last expression is bounded by

$$f(x_{k,r}^n) \leq \left(\frac{k}{n-1}\right) f(x_{k-1,r}^{n-1}) + \left(1 - \frac{k}{n-1}\right) f(x_{k,r}^{n-1}), \quad 1 \leq k \leq n - 1.$$

Hence, summing it follows that

$$(n - 1) \sum_{k=1}^{n-1} f(x_{k,r}^n) \leq \sum_{k=1}^{n-1} k \, f(x_{k-1,r}^{n-1}) + \sum_{k=1}^{n-1} (n - 1 - k) f(x_{k,r}^{n-1}),$$

where changing the limits of summation in the first sum gives that the sum is equal to

$$\sum_{k=0}^{n-2} (k + 1) \, f(x_{k,r}^{n-1}) + \sum_{k=1}^{n-1} (n - 1 - k) f(x_{k,r}^{n-1})$$

$$= f(x_{0,r}^{n-1}) + \sum_{k=1}^{n-2}(k+1)\, f(x_{k,r}^{n-1}) + \sum_{k=1}^{n-2}(n-1-k) f(x_{k,r}^{n-1}).$$

Whence, adding $(n-1) f(x_{0,r}^{n}) = (n-1) f(x_{0,r}^{n-1})$ to both sides (since $x_{0,r}^{n-1} = x_{0,r}^{n} = 0$), it follows that the sum equals

$$(n-1)\sum_{k=0}^{n-1} f(x_{k,r}^{n}) \leq n f(x_{0,r}^{n-1}) + n \sum_{k=1}^{n-2} f(x_{k,r}^{n-1}) = n \sum_{k=0}^{n-2} f(x_{k,r}^{n-1}),$$

and, consequently, $S_L(f, n) \leq S_L(f, n-1)$, as we wanted to prove in this case.

When f is increasing and concave, we apply the above result to $-f$. The proof is thus finished. □

Now, when the double sequence $\{r_k^n\}$ depends only on the level n, then $\beta_k^n = \beta_n = r_n - r_{n-1}$, which is negative when $\{r_n\}$ is decreasing. In particular, it holds for $r_n = r/n$, all n.

Borwein, Borwein, and Sims used a digitally assisted approach to address the question of the monotone convergence of the Riemann sums of f, [10]. By *digital assistance* , we mean the use of mathematical computer packages—symbolic, numeric, geometric, or graphical, such as Maple and Mathematica; specialized packages or general purpose languages such as Fortran, C++; and web applications and databases including Google, MathSciNet, ArXiv, and others. In this instance, digital assistance leads the authors through Google to the work of András [3], the evaluation of arctan(1), and the consideration of the sequences $\{x_n\}$, $\{y_n\}$, defined by

$$x_n = \sum_{k=1}^{n} \frac{n}{n^2+k^2}, \quad \text{and} \quad y_n = \sum_{k=0}^{n-1} \frac{n}{n^2+k^2}, \quad n = 1, 2, \ldots$$

These sequences can be expressed as $\{A_{n+1}(f)\}$ and $\{B_{n-1}(f)\}$ for the function $f(x) = 1/(1+x^2)$ and converge to the integral of f on $[0, 1]$. In fact, $\{x_n\}$ increases with n, whereas the monotonicity of $\{y_n\}$ remains an open question [10]. Now, f is decreasing on $[0, 1]$ and has an inflection point at $\sqrt{3}/3$, where it changes from concave to convex. The function $f = \chi_{[0,1/2]}$, which exhibits similar characteristics, shows that no monotonicity result is viable under these conditions [10]. However, a result is available if convex and concave are interchanged. In that case, we have:

Theorem 12 *Let f be a decreasing function defined on $[0, 1]$ such that f is convex on $[0, c]$ and concave on $[c, 1]$. Then $\{S_R(f, \mathcal{P}_n)\}$ increases and $\{S_L(f, \mathcal{P}_n)\}$ decreases with n.*

And, if f is increasing on $[0, 1]$, concave on $[0, c]$, and convex on $[c, 1]$, then $\{S_R(f, \mathcal{P}_n)\}$ decreases and $\{S_L(f, \mathcal{P}_n)\}$ increases with n.

Proof Assume first that f is decreasing, convex on $[0, c]$, and concave on $[c, 1]$, and let

$$g(x) = \begin{cases} f(x), & 0 \leq x < c, \\ f(c), & c \leq x \leq 1, \end{cases} \quad \text{and,} \quad h(x) = \begin{cases} f(c), & 0 \leq x < c, \\ f(x), & c \leq x \leq 1. \end{cases}$$

We claim that g is nonincreasing and convex on $[0, 1]$. Clearly, g is nonincreasing. As for convexity, it suffices to verify (25) for $0 \leq x < c < y \leq 1$, and $0 < \alpha < 1$. Note that then $\alpha x + (1 - \alpha) c \leq c \leq y$. Hence,

$$g(\alpha x + (1 - \alpha)y) \leq g(\alpha x + (1 - \alpha)c) = f(\alpha x + (1 - \alpha)c)$$

$$\leq \alpha f(x) + (1 - \alpha) f(c) = \alpha g(x) + (1 - \alpha) g(y).$$

Likewise, h is concave and decreasing on $[0, 1]$. Then, by Theorem 8, $\{S_R(g, \mathcal{P}_n)\}$ and $\{S_R(h, \mathcal{P}_n)\}$ increase with n, while $\{S_L(g, \mathcal{P}_n)\}$ and $\{S_L(h, \mathcal{P}_n)\}$ decrease with n. Moreover, since $g(x) + h(x) = f(x) + f(c)$, it follows that $S_R(g, \mathcal{P}_n) + S_R(h, \mathcal{P}_n) = S_R(f, \mathcal{P}_n) + S_R(f(c), \mathcal{P}_n) = S_R(f, \mathcal{P}_n) + f(c)$ increases with n, and so $\{S_R(f, \mathcal{P}_n)\}$ increases with n.

Similarly, the relation $S_L(g, \mathcal{P}_n) + S_L(h, \mathcal{P}_n) = S_L(f, \mathcal{P}_n) + f(c)$ yields that $\{S_L(f, \mathcal{P}_n)\}$ decreases with n, and we are done in this case. The proof of the second statement follows by applying the first part to $-f$.   □

A similar argument to Theorem 12 gives the following conclusion [10]. If a function f defined on I is concave on $(0, 1)$ and symmetric about the midpoint $x = 1/2$, then the sequence with terms

$$S_R(f, \mathcal{P}_n) - \frac{f(1/2) - f(0)}{n}$$

increases as n increases. The proof is left to the interested reader. In fact, along with convexity, the notions of symmetry and symmetrization play an important role in this area [10].

Chapter 3
A Convergence Theorem

In this chapter, we prove a basic convergence theorem for the Riemann integral. Although the Riemann integral does not enjoy satisfactory properties with respect to pointwise limits, Theorem 13 allows for the taking of limits. Our general setting is as follows: Let $\{\psi_n\}$ be a bounded family of Riemann integrable functions such that $\lim_n \int_I \psi_n \varphi = \alpha \int_I \varphi$ for all φ in a family of Riemann integrable functions whose span, i.e., the finite linear combinations of these functions, approximates the Riemann integrable functions on I. Then, $\lim_n \int_I \psi_n f = \alpha \int_I f$ for every Riemann integrable function f on I.

Now, since by (4) above, given $\varepsilon > 0$, there are (piecewise constant) *step functions* u and U, say, i.e., finite linear combinations of characteristic functions of subintervals J of I such that $u \leq f \leq U$ and $\int_I (U - u) < \varepsilon$, the collection of the characteristic functions of subintervals of I is one such family. And, graphically or otherwise, from this, it follows that, given $\varepsilon > 0$, there are continuous functions g, G, say, such that $g(x) \leq f(x) \leq G(x)$ throughout I, and $\int_I (G - g) \leq \varepsilon$. We may then invoke the Weierstrass approximation theorems (proven below) to the effect that algebraic and trigonometric polynomials approach continuous functions uniformly on I and, therefore, incorporate the algebraic and trigonometric monomials as families the convergence theorem applies to.

We prove the following basic result:

Theorem 13 *Let Π be an admissible family of partitions of a finite interval $I \subset \mathbb{R}$, and suppose that $\{\psi_n\}$ is a uniformly bounded sequence of Riemann integrable functions on I such that for some $\alpha \in \mathbb{R}$ and all subintervals $J \subset I$ that belong to some partition $\mathcal{P}$ of I in Π,*

$$\lim_n \int_J \psi_n = \alpha \, |J|. \tag{37}$$

© The Author(s), under exclusive license to Springer Nature Switzerland AG 2022
A. Torchinsky, *A Modern View of the Riemann Integral*,
Lecture Notes in Mathematics 2309, https://doi.org/10.1007/978-3-031-11799-2_3

Then, for every Riemann integrable function f on I,

$$\lim_n \int_I f \, \psi_n = \alpha \int_I f. \tag{38}$$

Furthermore, let $\mathcal{F} = \{\varphi\}$ be a family of Riemann integrable functions on I so that continuous functions on I can be approximated uniformly by functions in the span of $\mathcal{F}$. Then, if $\{\psi_n\}$ is a bounded family of Riemann integrable functions such that $\lim_n \int_I \psi_n \varphi = \alpha \int_I \varphi$, for all $\varphi \in \mathcal{F}$, it follows that $\lim_n \int_I \psi_n f = \alpha \int_I f$ for every Riemann integrable function f on I.

Proof By Theorem 1, the Riemann integral coincides with the Π-Darboux integral on I, so it suffices to prove (38) for functions f that are Π-Darboux integrable on I. Now, given $\varepsilon > 0$, let the step functions u and U be finite linear combinations of characteristic functions of subintervals J of I that belong to a partition $\mathcal{P}$ of I in Π, such that $u \leq f \leq U$ and $\int_I (U - u) < \varepsilon$. Moreover, since $|f(x) - u(x)| \leq U(x) - u(x)$ throughout I, given $\varepsilon > 0$, there is a step function u (based on a partition in Π), such that

$$\left| \int_I f - \int_I u \right| \leq \int_I |f - u| \leq \varepsilon. \tag{39}$$

Suppose first that $\alpha = 0$. Then, given $\varepsilon > 0$, pick u that corresponds to $\varepsilon/2M$ in (39). Then,

$$\int_I f \psi_n = \int_I (f - u) \, \psi_n + \int_I u \, \psi_n,$$

where the first summand on the right-hand side is bounded by $M \varepsilon/2M = \varepsilon/2$, uniformly in n. Now, if u is a step function based on a partition $\mathcal{P}$ of I in Π, by linearity, (37) yields $\lim_n \int_I u \, \psi_n = 0$, and, consequently, the second summand does not exceed $\varepsilon/2$ for all sufficiently large n, and so combining these observations $|\int_I f \psi_n| \leq \varepsilon$ for all sufficiently large n, and (38) holds.

On the other hand, when $\alpha \neq 0$, pick u corresponding to $\varepsilon/3(M + |\alpha|)$ in (39), and write

$$\int_I f \psi_n - \alpha \int_I f = \int_I (f - u) \, \psi_n$$

$$+ \left(\int_I u \, \psi_n - \alpha \int_I u \right) + \left(\alpha \int_I u - \alpha \int_I f \right),$$

where the first summand on the right-hand side is bounded by $M\varepsilon/3(M+|\alpha|) \leq \varepsilon/3$ uniformly in n, and the third summand by $|\alpha| \, \varepsilon/3(M + |\alpha|) \leq \varepsilon/3$ independently of n. As for the second summand, it is estimated as above and does not exceed $\varepsilon/3$ for sufficiently large n. Thus, (38) holds, and the proof is finished in this case.

As for the second statement, given $\varepsilon > 0$, let g, G be continuous functions on I such that $g(x) \leq f(x) \leq G(x)$ throughout I, and $\int_I (G - g) \leq \varepsilon/2$. Now, since g is continuous, there exists Φ in the span of $\mathcal{F}$ such that $|g(x) - \Phi(x)| \leq \varepsilon/2, x \in I$, and therefore,

$$|f(x) - \Phi(x)| \leq |f(x) - g(x)| + |g(x) - \Phi(x)| \leq \varepsilon, \quad x \in I.$$

The argument is now completed by repeating the above proof with Φ in place of u in (39) and after.   □

3.1   The Riemann–Lebesgue Lemma

The Riemann–Lebesgue lemma asserts that the Fourier coefficients of an integrable function tend to 0 as $n \to \infty$. Picking Π as the family of all partitions of $I = [-\pi, \pi]$ and $\psi_n(x) = \sin(nx), \cos(nx)$, Theorem 13 gives the lemma for Riemann integrable functions.

The results that follow are variations on this theme. There are instances where the admissible family is Π_e, [21]. Suppose that $\{\psi_n\}$ is a bounded sequence of Riemann integrable functions on $I = [0, 1]$ such that $\int_{I_k^n} \psi_n = 0$ for all $I_k^n \in \mathcal{P}_n$, all n. Then, we claim that for each Riemann integrable function f on I, we have $\lim_n \int_I f \psi_n = 0$. This readily follows from Theorem 13. Indeed, it suffices to verify that (37) holds for $J = [0, b]$ with $0 < b \leq 1$, for if this is the case, since $\chi_{[a,b]} = \chi_{[0,b]} - \chi_{[0,a)}$, it also holds for an arbitrary subinterval of I.

Observe that for n with $1/n < b$, an interval in $\mathcal{P}_n$ is contained in J or is disjoint with J, or, at most one, say $I_{k_0}^n$, straddles J. Then

$$\int_J \psi_n = \sum_{k=1}^{k_0-1} \int_{I_k^n} \psi_n + \int_{J \cap I_{k_0}^n} \psi_n = \int_{J \cap I_{k_0}^n} \psi_n,$$

and, consequently, with M a bound for the $\{\psi_n\}$,

$$\left| \int_J \psi_n \right| \leq M |J \cap I_{k_0}^n| \leq M/n \to 0, \quad \text{as } n \to \infty.$$

This observation applies in the following context. Let $\{h_n\}$ be a uniformly bounded sequence of Riemann integrable odd functions defined on $[-1, 1]$, such that $\psi_n(x) = h_n(\sin(2\pi nx))$ is Riemann integrable for all n. Then, by symmetry $\int_{I_k^n} \psi_n = 0$ for all $I_k^n \in \mathcal{P}_n$, and so $\lim_n \int_I f \psi_n = 0$ for all integrable functions f. A similar conclusion holds for $\psi_n(x) = h_n(\cos(\pi nx))$.

Theorem 13 also gives the following result in the spirit of the Riemann–Lebesgue lemma [4, 51, 99]:

Proposition 12 *Let f be a Riemann integrable function defined on $I = [a, b]$, $a \geq 0$, and let ψ, φ be defined for $x \geq 0$, and Riemann integrable on finite subintervals of $\mathbb{R}$. Then:*

(i) *If $\lim_{x \to \infty} (1/x) \int_{[0,x]} \psi = \alpha$, $\lim_n \int_I f(x)\psi(nx)\,dx = \alpha \int_I f$.*

(ii) *If $\psi(x + \beta) = \psi(x)$ for some $\beta > 0$ and all $x \in \mathbb{R}$,*

$$\lim_n \int_I f(x)\,\psi(nx)\,dx = \left(\frac{1}{\beta} \int_{[0,\beta]} \psi \right) \int_I f.$$

(iii) *If $\varphi(x + \beta) = -\varphi(x)$ for some $\beta > 0$ and all $x \in \mathbb{R}$,*

$$\lim_n \int_I f(x)\,\varphi(nx)\,dx = 0.$$

Proof Note that if $\psi_n(x) = \psi(nx)$, $\{\psi_n\}$ is a uniformly bounded sequence of Riemann integrable functions on I; similarly for $\varphi_n(x) = \varphi(nx)$. To prove (i), by Theorem 13, it suffices to verify that for all subintervals $J \subset I$,

$$\lim_n \int_J \psi(nx)\,dx = \alpha |J|. \tag{40}$$

First let $J = [a, c]$, where $a < c \leq b$. Then,

$$\int_J \psi(nx)\,dx = \frac{1}{n} \int_{[an,cn]} \psi = c \left(\frac{1}{cn} \int_{[0,cn]} \psi \right) - a \left(\frac{1}{an} \int_{[0,an]} \psi \right),$$

which by assumption tends to $\alpha c - \alpha a = \alpha |J|$ as $n \to \infty$. Next, if $J = [c, d]$ is an arbitrary subinterval of I, $\int_J \psi(nx)\,dx = \int_{[a,d]} \psi(nx)\,dx - \int_{[a,c]} \psi(nx)\,dx$, and, consequently, $\lim_n \int_J \psi(nx)\,dx = \alpha(d - a) - \alpha(c - a) = \alpha(d - c) = \alpha|J|$, (40) holds, and the proof of (i) is finished.

Now, if ψ is periodic, for $x > 0$, write $x = k\beta + \gamma$, where k is an integer and $0 \leq \gamma < \beta$. Note that $k \to \infty$ as $x \to \infty$, and $\lim_{x \to \infty} k/x = 1/\beta$.

Then, by periodicity, $\int_{[0,x]} \psi = k \int_{[0,\beta]} \psi + \int_{[k\beta,\gamma]} \psi$, where with M_ψ a bound for ψ,

$$\left| \frac{1}{x} \int_{[k\beta,\gamma]} \psi \right| \leq \frac{1}{x} \beta M_\psi \to 0, \quad \text{as } x \to \infty.$$

Therefore,

$$\lim_{x \to \infty} \frac{1}{x} \int_{[0,x]} \psi = \lim_{x \to \infty} \frac{1}{x} k \int_{[0,\beta]} \psi + \lim_{x \to \infty} \frac{1}{x} \int_{[k\beta,\gamma]} \psi = \frac{1}{\beta} \int_{[0,\beta]} \psi,$$

and by (i) we are done.

As for (iii), note that since $\varphi(x + 2\beta) = -\varphi(x + \beta) = \varphi(x)$, φ is periodic of period 2β, with

$$\int_{[0,2\beta]} \varphi = \int_{[0,\beta]} \varphi + \int_{[\beta,2\beta]} \varphi = -\int_{[0,\beta]} \varphi(x + \beta)\, dx + \int_{[\beta,2\beta]} \varphi$$

$$= -\int_{[\beta,2\beta]} \varphi + \int_{[\beta,2\beta]} \varphi = 0,$$

and, therefore, (ii) holds in this case with $\alpha = 0$. This gives (iii) and completes the proof. □

A particular instance of (ii) above, which includes the Riemann–Lebesgue lemma, can be proved directly as follows. Let f, ψ be Riemann integrable on $I = [0, 1]$, where ψ is periodic of period $\beta = 1$. Observe that with $I_k^n = [(k-1)/n, k/n]$,

$$\int_I f(x)\, \psi(nx)\, dx = \sum_{k=1}^{n} \int_{I_k^n} f(x)\, \psi(nx)\, dx,$$

where

$$\int_{I_k^n} f(x)\, \psi(nx)\, dx = \frac{1}{n} \int_I f\big((k-1)/n + y/n\big)\, \psi(y)\, dy$$

$$= \frac{1}{n} \int_I \Big(f\big((k-1)/n + y/n\big) - f\big((k-1)/n\big) \Big) \psi(y)\, dy$$

$$+ f\big((k-1)/n\big) \frac{1}{n} \int_I \psi(y)\, dy,$$

and, consequently, rearranging and summing,

$$\int_I f(x)\, \psi(nx)\, dx - \left(\int_I \psi(y)\, dy \right) S_L(f, \mathcal{P}_n)$$

$$= \int_I \frac{1}{n} \sum_{k=1}^{n} \Big(f\big((k-1)/n + y/n\big) - f\big((k-1)/n\big) \Big) \psi(y)\, dy.$$

Now, the right-hand side above can be estimated in various ways, depending on the properties of f. When f is Riemann integrable on I, with M_ψ a bound for ψ, it is bounded by

$$M_\psi \frac{1}{n} \sum_{k=1}^{n} \text{osc}\,(f, I_k^n) \to 0, \quad \text{as } n \to \infty,$$

and, therefore, since $\lim_n S_L(f, \mathcal{P}_n) = \int_I f$, we conclude that (ii) holds.

Also, if f satisfies a Lipschitz condition of order α, then as in (17), osc $(f, I_k^n) \leq L n^{-\alpha}$, and $\int_I f(x) \, \psi(nx) \, dx - \left(\int_I \psi \right) \left(\int_I f \right) = O(n^{-\alpha})$. Thus, when $\int_I \psi = 0$, this coincides with the rate of decay of the Fourier coefficients of functions that satisfy a Lipschitz condition of order α.

If f is continuous on I, it can be estimated by

$$M_\psi \, \omega_f(1/n).$$

And, if f is differentiable with f' Riemann integrable,

$$f\big((k-1)/n + y/n\big) - f\big((k-1)/n\big) = \frac{1}{n} f'(c_k^n),$$

with c_k^n depending on y, and so,

$$\left| \sum_{k=1}^{n} \Big(f\big((k-1)/n + y/n\big) - f\big((k-1)/n\big) \Big) - \frac{1}{n} \sum_{k=1}^{n} f'\big((k-1)/n\big) \right|$$

$$\leq \frac{1}{n} \sum_{k=1}^{n} \operatorname{osc}(f', I_k^n).$$

Therefore,

$$n \left| \int_I f(x) \, \psi(nx) \, dx - \left(\int_I \psi \right) S_L(f, \mathcal{P}_n) - S_L(f', \mathcal{P}_n) \right|$$

$$\leq \frac{1}{n} \sum_{k=1}^{n} \operatorname{osc}(f', I_k^n) \to 0, \quad \text{as } n \to \infty,$$

and the rate of approximation is better than $1/n$.

The results of Proposition 12 also hold for a continuous parameter λ going to infinity in place of the n. In this context, we summarize the above results as follows:

Proposition 13 *Let ψ be Riemann integrable on every $I = [a, b]$. Then the following are equivalent:*

(i)

$$\lim_{\lambda \to \infty} \frac{1}{\lambda} \int_{[0,\lambda]} \psi = 0.$$

(ii) *For each Riemann integrable function f on I,*

$$\lim_{\lambda \to \infty} \int_I f(x) \, \psi(\lambda x) \, dx = 0.$$

(iii) *For a fixed $c \in (0, \infty)$,*

$$\lim_{\lambda \to \infty} \int_{[0,c]} \psi(\lambda x) \, dx = 0.$$

Proof (i) implies (ii) follows from Proposition 12. By considering $f = \chi_{[0,c]}$, we see that (ii) implies (iii). As for (iii) implies (i), observe that for each $\lambda \in (0, \infty)$, we have

$$\int_{[0,c]} \psi(\lambda x) \, dx = \frac{1}{\lambda} \int_{[0,c\lambda]} \psi = c \frac{1}{c\lambda} \int_{[0,c\lambda]} \psi.$$

This completes the proof. $\qquad\qquad\qquad\qquad\qquad\qquad\qquad\qquad\qquad\qquad$ $\square$

An improved version of the theorem actually holds for monotone functions and therefore also for functions of bounded variation [102]. In this context, we have:

Proposition 14 *Let $I = [a, b]$ be a closed bounded interval in $\mathbb{R}$, let f, ψ be defined on I such that f is monotone increasing, ψ is continuous, and $\Psi(x) = \Psi(a) + \int_{[a,x]} \psi$, the antiderivative of ψ in $\mathbb{R}$, is bounded and integrable on bounded intervals. Then*

$$\int_I f(x) \psi(\lambda x) \, dx = O(1/\lambda).$$

Proof Let $\{I_k^n\}$ be a partition of I in Π_e. Then, for $\lambda > 0$, by the mean value theorem, for each $k = 1, 2, \ldots, n$, there is a $c_k^n \in I_k^n$ such that

$$\Psi(\lambda x_k^n) - \Psi(\lambda x_{k-1}^n) = \lambda \psi(\lambda c_k^n)(x_k^n - x_{k-1}^n), \quad 1 \le k \le n. \qquad (41)$$

Consider then the Riemann sum

$$S_{\Pi_e}(f\psi(\lambda \cdot), \mathcal{P}_n, C^n) = \sum_{k=1}^n f(c_k^n) \, \psi(\lambda c_k^n) \left(x_k^n - x_{k-1}^n\right)$$

with tags $C^n = \{c_k^n\}$, which, by virtue of (41), can be written as

$$S_{\Pi_e}(f\psi(\lambda \cdot), \mathcal{P}_n, C^n) = \frac{1}{\lambda} \sum_{k=1}^n f(c_k^n) \left(\Psi(\lambda c_k^n) - \Psi(\lambda c_{k-1}^n)\right).$$

Now, summing by parts, it follows that

$$S_{\Pi_e}(f\psi(\lambda \cdot), \mathcal{P}_n, C^n) = \frac{1}{\lambda} \left(f(c_n^n) \Psi(\lambda) - f(c_1^n) \Psi(0)\right)$$

$$+\frac{1}{\lambda}\sum_{k=1}^{n-1}\left(f(c_k^n)-f(c_{k-1}^n)\right)\Psi(\lambda c_k^n),$$

where the first summand is $O(1/\lambda)$ on account of the assumptions, and the second, with M_Ψ a bound for Ψ, is bounded by

$$M_\Psi\left(f(1)-f(0)\right)\frac{1}{\lambda} = O(1/\lambda),$$

and the proof is finished. $\square$

3.2   The Weierstrass Approximation Theorems

A key ingredient in the proof of the convergence theorem is the fact that the linear span of the characteristic functions of the subintervals of I contains the step functions on I, which approximate the Riemann integrable functions on I. The continuous functions on I share this property. The Weierstrass approximation theorems give the uniform approximation of continuous functions by algebraic and trigonometric polynomials on I, and this in turn allows for the algebraic and trigonometric monomials to play the role of the step functions in Theorem 13.

We will begin by considering the approximation by algebraic polynomials. Of the various ways to approach this result, we adopt the probability inspired proof of Sergei Bernstein. Given a bounded function f on $I = [0, 1]$, for $n \geq 1$, let the *Bernstein polynomial* $B_n(f)(x)$ associated to f, be the polynomial of degree $\leq n$, given by

$$B_n(f)(x) = \sum_{k=0}^{n} f(k/n)\binom{n}{k}x^k(1-x)^{n-k}, \quad x \in [0, 1].$$

Note that in general the Bernstein polynomial $B_n(p)(x)$ of a polynomial p is not $p(x)$, e.g., for $p(x) = x^2$, $B_2(p)(x) = x(1+x)/2$, and, more generally, $B_n(p)(x) = x^2 + x(1-x)/n$.

Whereas Bernstein's original proof makes use of basic facts concerning the binomial distribution and Chebyshev's theorem, the proof presented here uses the identity

$$n\,x\,(1-x) = \sum_{k=0}^{n}(nx-k)^2\binom{n}{k}x^k(1-x)^{n-k}, \quad x \in [0, 1]. \tag{42}$$

To see this first, observe that by the binomial formula, for $x, y \in [0, 1]$,

$$\left(y + (1 - x)\right)^n = \sum_{k=0}^{n} \binom{n}{k} y^k (1 - x)^{n-k}, \tag{43}$$

and, consequently, setting $y = x$, it follows that

$$1 = \left(x + (1 - x)\right)^n = \sum_{k=0}^{n} \binom{n}{k} x^k (1 - x)^{n-k}. \tag{44}$$

Now, differentiating with respect to y in (43), setting $y = x$, and multiplying through by x yield

$$nx = \sum_{k=0}^{n} \binom{n}{k} k x^k (1 - x)^{n-k}. \tag{45}$$

And, differentiating (43) with respect to y twice, setting $y = x$, and multiplying through by x^2 give

$$n(n - 1)x^2 = \sum_{k=0}^{n} \binom{n}{k} k(k - 1)x^k (1 - x)^{n-k}. \tag{46}$$

Finally, multiplying the identity (44) by $n^2 x^2$, subtracting the identity (45) times $(2nx - 1)$, and adding the relation (46) give (42).

We are now ready to proceed with the proof:

Approximation by Algebraic Polynomials *Let f be a bounded function defined on $I = [0, 1]$. Then, $\lim_n B_n(f)(x) = f(x)$ at each point of continuity $x \in I$ of f. Furthermore, if f is continuous on I, the convergence is uniform.*

Proof Let x be a point of continuity of f. Then, given $\varepsilon > 0$, there is $\delta > 0$ such that $|f(x) - f(y)| \le \varepsilon/2$ if $|x - y| \le \delta$. Now, by (44), we have

$$f(x) - B_n(f)(x) = \sum_{k=0}^{n} \left(f(x) - f(k/n)\right) \binom{n}{k} x^k (1 - x)^{n-k}.$$

We divide the sum on the right-hand side above into two parts. Let $K_1 = \{k : 0 \le k \le n,$ and $|x - k/n| \le \delta\}$, and, $K_2 = \{0, \dots, n\} \setminus K_1$. Then, by the continuity of f at x and (44),

$$\left| \sum_{k \in K_1} \left(f(x) - f(k/n)\right) \binom{n}{k} x^k (1 - x)^{n-k} \right|$$

$$\leq \frac{\varepsilon}{2} \sum_{k \in K_1} \binom{n}{k} x^k (1-x)^{n-k} \leq \frac{\varepsilon}{2} \sum_{k=0}^{n} \binom{n}{k} x^k (1-x)^{n-k} = \frac{\varepsilon}{2}.$$

Furthermore, since for $k \in K_2$, $(nx - k)^2 > n^2\delta^2$, with M a bound for f in I, on account of (42), we have

$$\left| \sum_{k \in K_2} \left(f(x) - f(k/n) \right) \binom{n}{k} x^k (1-x)^{n-k} \right|$$

$$\leq 2M \sum_{k \in K_2} \frac{(nx - k)^2}{n^2\delta^2} \binom{n}{k} x^k (1-x)^{n-k}$$

$$\leq \frac{2M}{n^2\delta^2} \, n \, x(1-x) \leq \frac{M}{2n\delta^2} \leq \frac{\varepsilon}{2},$$

provided that $n \geq n_0$, where n_0 is an integer $\geq M/\varepsilon\delta^2$.

Thus, combining the above estimates, it follows that $|f(x) - B_n(f)(x)| \leq \varepsilon$, for all $n \geq n_0$, and the convergence at a point x of continuity of f has been established.

Finally, if f is continuous on I, it is uniformly continuous there, and given $\varepsilon > 0$, there is $\delta > 0$ such that $|f(y) - f(z)| \leq \varepsilon/2$ if $|y - z| \leq \delta$, i.e., the same δ works for all the points in I. Hence, the same n_0 works for all $x \in I$, and the proof is finished.

$\square$

Note that if the approximation theorem holds for functions defined on $[0, 1]$, it also holds for functions defined on an arbitrary interval $[a, b]$, and conversely. Since the proof of both assertions is similar, we only do the first. Given a continuous function f defined on $[a, b]$, the function $g(x) = f(a + (b - a)x)$ is a continuous function on $[0, 1]$ such that $\max_{0 \leq x \leq 1} |g(x)| = \max_{a \leq t \leq b} |f(t)|$. Then, given $\varepsilon > 0$, let p be a polynomial such that

$$\max_{x \in I} |g(x) - p(x)| \leq \varepsilon, \quad \text{or,} \quad \max_{x \in I} |f(a + (b - a)x) - p(x)| \leq \varepsilon.$$

Finally, since $p(x)$ is a polynomial in x, $q(t) = p\big((t-a)/(b-a)\big)$ is a polynomial in t that satisfies

$$\max_{a \leq t \leq b} \left| f(t) - q(t) \right| \leq \varepsilon,$$

as we wanted to show.

As for arbitrary Riemann integrable functions f on I, we can say something about the limits of oscillation of $\{B_n(f)(x)\}$, the $\lim_n \int_I B_n$, and functions with vanishing moments.

Given $\varepsilon > 0$, let g, G be continuous functions such that $g(x) \leq f(x) \leq G(x)$ throughout I and $\int_I (G - g) \leq \varepsilon$. By the monotonicity property of the Bernstein polynomials, $B_n(g)(x) \le B_n(f)(x) \le B_n(G)(x)$ throughout I, and, consequently,

$$\limsup_n B_n(f)(x) - \liminf_n B_n(f)(x) \le G(x) - g(x), \quad x \in I.$$

This statement implicitly states the fact that $\lim_n B_n(f)(x) = f(x)$ at the points x of continuity of f because at those points we may chose g, G with $g(x) = G(x)$. Hence, by Proposition 2, $B_n(f)(x)$ converges pointwise to $f(x)$ at infinitely many points of I, but not uniformly. Yet we may take a limit. Indeed, since

$$\int_I \binom{n}{k} x^k (1-x)^{n-k} = \frac{1}{n+1}, \quad n = 0, 1, 2, \ldots,$$

it readily follows that

$$\int_I B_n(f)(x)\,dx = \frac{1}{n+1} \sum_{n=0}^{n} f(k/n)$$

is a Riemann sum of f, and, consequently,

$$\lim_n \int_I B_n(f)(x)\,dx = \int_I f.$$

We can also identify those Riemann integrable functions f on an interval $I = [a, b]$ with vanishing moments of all orders there, i.e.,

$$\int_I f(x) x^n\,dx = 0, \quad n = 0, 1, 2, \ldots$$

Note that for such a function f, by linearity, we have $\int_I f\,p = 0$ for any polynomial p, and so, if h is a continuous function on I, it follows that $\int_I f\,h = \int_I f\,(h - p)$. Then, given $\varepsilon > 0$, pick p so that $|h(x) - p(x)| \le \varepsilon$ throughout I, and so $\left| \int_I f\,h \right| \le \varepsilon \int_I |f|$, which implies that $\int_I f h = 0$. Now, with g, G as above and M_f a bound for f, we have

$$\int_I f^2 = \int_I f(f - G) \le M_f \int_I (G - f) \le M_f \int_I (G - g) \le M_f \varepsilon,$$

And so, $\int_I f^2 = 0$. It then readily follows that $f(x) = 0$ at each point x of continuity of f, and so, when f is continuous on I, we have that f^2, and hence, f vanishes on I. On the other hand, χ_C, the characteristic function of the Cantor set C, satisfies $\int_{[0,1]} \chi_C^2 = 0$, so nothing else can be said about arbitrary integrable functions.

We will discuss next the approximation by trigonometric polynomials. Traditionally, one works in the interval $[-\pi, \pi]$, but the results can be transferred to any interval $[a, b]$, [77]. By a *trigonometric polynomial* $t_N(x)$ of degree N on $[-\pi, \pi]$, we mean an expression of the form

$$t_N(x) = \frac{a_0}{2} + \sum_{k=1}^{N} a_k \cos(kx) + b_k \sin(kx), \quad x \in [-\pi, \pi].$$

It turns out that trigonometric polynomials approximate continuous periodic functions uniformly on $[-\pi, \pi]$ iff algebraic polynomials do so [77]. Nevertheless, since the original proof of this result by the septuagenarian Weierstrass involves the important notion of approximate identity, we prefer to prove it directly. An efficient approach is via the polynomials introduced by the nineteen years old Leopold Fejér, but some of their properties are not apparent at first sight without some knowledge of Fourier series. This is a spot where a proof for a mathematician in a hurry, as J. E. Littlewood used to say, is called for. So, in this setting, we have:

Proposition 15 *For* $\lambda > 0$, *consider the expression* $F(x, \lambda)$ *defined on* $[-\pi, \pi]$ *by*

$$F(x, \lambda) = \frac{1 - \cos(\lambda x)}{1 - \cos(x)}.$$

Then, $F(x, n)$ *is a trigonometric polynomial, for all integers* $n = 1, 2, \ldots$
Furthermore, if for positive real numbers α, λ, *with* $\alpha < 3$, *we let*

$$I_{\alpha, \lambda} = \frac{1}{\lambda^3} \int_{[-\pi, \pi]} |x|^{\alpha} F(x, \lambda)^2 \, dx, \quad I_{0, \lambda} = I_{\lambda},$$

there are positive constants c, C *independent of* λ *such that* $c \le I_{\lambda} \le C$, *and a constant* $C_{\alpha} = O\left(1/(3 - \alpha)\right)$ *independent of* λ, *such that* $I_{\alpha, \lambda} \le C_{\alpha} \lambda^{-\alpha}$.

Proof The first assertion is readily verified by induction when the degree of the polynomial is equal to 2^k; this often suffices for applications. Indeed, since

$$1 - \cos(2x) = 2 \sin^2(x) = 2(1 - \cos(x))(1 + \cos(x)),$$

it follows that $F(x, 2) = 2(1 + \cos(x))$ is a trigonometric polynomial when $\lambda = 2k$ with $k = 1$. And, for $k > 1$,

$$F(x, 2^k) = \frac{1 - \cos(2^k x)}{1 - \cos(2^{k-1} x)} F(x, 2^{k-1}),$$

where the first factor is the trigonometric polynomial $2\left(1 + \cos(2^{k-1}x)\right)$, and by the inductive assumption, the second factor is also a trigonometric polynomial. Since the product of trigonometric polynomials is again a trigonometric polynomial, the proof is complete in this particular case.

To verify directly that $F(x, n)$ is a trigonometric polynomial for all positive integers n, note that telescoping,

$$1 - \cos(nx) = \sum_{k=0}^{n-1} \big(\cos(kx) - \cos((k+1)x)\big).$$

Now, the summand corresponding to $k = 0$ is equal to $1 - \cos(x)$, and for $k = 1, \ldots, n-1$, we have

$$\cos(kx) - \cos((k+1)x)$$
$$= (1 - \cos(x)) - \sum_{j=1}^{k} \big(\cos((j-1)x) - 2\cos(jx) + \cos((j+1)x)\big),$$

where each summand is equal to

$$\big(\cos((j-1)x) + \cos((j+1)x)\big) - 2\cos(jx)$$
$$= 2\cos(jx)\cos(x) - 2\cos(jx) = 2\cos(jx)\big(\cos(x) - 1\big).$$

Hence,

$$1 - \cos(nx) = (1 - \cos(x)) \sum_{k=0}^{n-1} \Big(1 + \sum_{j=1}^{k} 2\cos(jx)\Big),$$

$$= (1 - \cos(x)) \Big(n + \sum_{j=1}^{n-1} 2(n-j)\cos(jx)\Big),$$

and $F(x, n)$ is a trigonometric polynomial of degree $(n-1)$, [49].

Next observe that since

$$1 - \cos(\lambda x) = \frac{1}{2}\sin^2(\lambda x/2), \quad \text{and,} \quad 1 - \cos(x) = \frac{1}{2}\sin^2(x/2),$$

we have

$$I_\lambda = \frac{2}{\lambda^3} \int_{[0,\pi]} \frac{\sin^4(\lambda x/2)}{\sin^4(x/2)}\, dx.$$

Let $\eta > 0$ be such that

$$1/2 \leq \frac{\sin(x)}{x} \leq 1, \quad 0 < x < \eta. \tag{47}$$

Then, since $\eta/\lambda \leq \pi$, it follows that

$$I_\lambda \geq \frac{2}{\lambda^3} \int_{[0,\eta/\lambda]} \frac{\sin^4(\lambda x/2)}{\sin^4(x/2)}\,dx.$$

Now, for x in the domain of integration above, we have $\lambda x/2 \leq (\eta/\lambda)(\lambda/2)$ $= \eta/2 \leq \eta$, and so from (47), it follows that

$$\sin^4(\lambda x/2) \geq \frac{1}{4^4}\lambda^4 x^4,$$

which, together with the estimate $1/\sin^4(x/2) \geq 2^4/x^4$, gives the lower bound

$$I_\lambda \geq \frac{2}{\lambda^3} \int_{[0,\eta/\lambda]} \frac{1}{4^4}\lambda^4 x^4 \frac{2^4}{x^4}\,dx = \frac{1}{2^3}\lambda \int_{[0,\eta/\lambda]} dx = \frac{1}{2^3}\eta.$$

As for the upper bound, the change of variables $\lambda x/2 = u$ gives

$$I_{\alpha,\lambda} = \frac{4}{\lambda^{4+\alpha}}2^\alpha \int_{[0,\pi\lambda/2]} u^\alpha \frac{\sin^4(u)}{\sin^4(u/\lambda)}\,du.$$

Since in the integral $u/\lambda \leq \pi/2$, and since $\sin(x)/x$ is decreasing for $0 < x < \pi/2$, we have

$$\frac{\sin(u/\lambda)}{(u/\lambda)} \geq \frac{2}{\pi}, \quad \text{and so,} \quad \frac{1}{\sin^4(u/\lambda)} \leq \frac{\pi^4}{2^4}\lambda^4 \frac{1}{u^4},$$

it readily follows that

$$I_{\alpha,\lambda} \leq \frac{4}{\lambda^{4+\alpha}}2^\alpha \frac{\pi^4}{2^4}\lambda^4 \int_{[0,\pi\lambda/2]} u^\alpha \frac{\sin^4(u)}{u^4}\,du$$

$$= \lambda^{-\alpha}\pi^4 2^{\alpha-2}\left(\int_{[0,1]} u^\alpha \frac{\sin^4(u)}{u^4}\,du + \int_{[1,\pi\lambda/2]} u^\alpha \frac{\sin^4(u)}{u^4}\,du \right).$$

Therefore, since the first integral above is $\leq 1/(1+\alpha)$, and

$$\int_{[1,\pi\lambda/2]} u^\alpha \frac{\sin^4(u)}{u^4}\,du \leq \int_{[1,\pi\lambda/2]} u^\alpha \frac{1}{u^4}\,du \leq 1/(3-\alpha), \quad \text{all large } \lambda,$$

we obtain the upper bound C for I_λ and the upper bound $C_\alpha \lambda^{-\alpha}$ for $I_{\alpha,\lambda}$, where for $\alpha \geq 0$, $C_\alpha = \pi^4 2^{\alpha-2}(1/(1+\alpha) + 1/(3-\alpha))$, and the proof is complete.   □

Concerning the expression $\int_I f(x - y)\,F(y,n)^2\,dy$ for n integer, where $I = [-\pi, \pi]$, since both factors in the (convolution) integral are periodic functions of period 2π, its value is unchanged if the interval of integration is replaced by any other interval of length 2π, and, in particular, by a change of variables, it is equal to $\int_I f(y) F(x - y, n)^2 \, dy$. Now, $F(x - y, n)$ is a trigonometric polynomial of order $(n-1)$ in $(x - y)$, and therefore, $F(x-y, n)^2$ is a trigonometric polynomial of order $2(n - 1)$ that can be expressed as a trigonometric sum of the same order in x, with coefficients that are trigonometric functions in y, and so $\int_I f(x - y) F(y, n)^2 \, dy$ is a trigonometric polynomial of order at most $2(n - 1)$ in x with coefficients that depend on f.

With this observation out of the way, we have:

Approximation by Trigonometric Polynomials *Let f be a periodic Riemann integrable function defined on $I = [-\pi, \pi]$, and let c_λ be such that $c_\lambda \int_I I_\lambda = 1$.*

 (i) *At each point of continuity $x \in I$ of f, we have*

$$\lim_{\lambda \to \infty} c_\lambda \int_I f(x - y) F(y, \lambda)^2 \, dy = f(x),$$

and if f is continuous, the convergence is uniform on I. Moreover, there is a sequence of trigonometric polynomials that converges uniformly to f.
 (ii) *If f is Lipschitz of order α on I with constant L, $0 < \alpha \leq 1$, we have*

$$\left| f(x) - c_\lambda \int_I f(x - y) F(y, \lambda)^2 \, dy \right| = O(1/\lambda^\alpha).$$

In particular, given an integer n, there are a trigonometric polynomial t_n of degree n and a constant C independent of f, such that

$$|f(x) - t_n(x)| \leq C L \frac{1}{n^\alpha}, \quad x \in I.$$

(iii) *If f is a continuous periodic function on I with modulus of continuity ω_f, given an integer n, there is a trigonometric polynomial t_n of degree n on I such that, for all real values of x,*

$$|f(x) - t_n(x)| = O\big(\omega_f(2\pi/n)\big).$$

Proof By Proposition 15, we may pick $c_\lambda \sim c\lambda^{-3}$ such that $c_\lambda \int_I F(y, \lambda)^2 \, dy = 1$, and for this value of c_λ, we have

$$f(x) - c_\lambda \int_I f(x - y) F(y, \lambda)^2 \, dy = c_\lambda \int_I \big(f(x) - f(x - y)\big) F(y, \lambda)^2 \, dy.$$

Now, if $x \in I$ is a point of continuity of f, given $\varepsilon > 0$, let $\delta > 0$ be such that $|f(x) - f(x - y)| \leq \varepsilon$ provided that $|y| \leq \delta$. Then,

$$\left| c_\lambda \int_I \left(f(x) - f(x - y) \right) F(y, \lambda)^2 \, dy \right|$$

$$\leq \varepsilon \, c_\lambda \int_{|y| \leq \delta} F(y, \lambda)^2 \, dy + c_\lambda \int_{\delta < |y| \leq \pi} \left| f(x) - f(x - y) \right| F(y, \lambda)^2 \, dy.$$

The first summand above is estimated by

$$\varepsilon \, c_\lambda \int_{|y| \leq \delta} F(y, \lambda)^2 \, dy \leq \varepsilon \, c_\lambda \int_{|y| \leq \pi} F(y, \lambda)^2 \, dy = \varepsilon.$$

As for the second summand, with $M = \sup |f|$, since

$$F(y, \lambda)^2 \leq \frac{4}{\left(1 - \cos(\delta) \right)^2}, \quad 0 < \delta \leq |x| \leq \pi,$$

it does not exceed

$$c_\lambda \int_{\delta < |y| < \pi} \left| f(x) - f(x - y) \right| F(y, \lambda)^2 \, dy \leq c_\lambda \frac{16 M \pi}{\left(1 - \cos(\delta) \right)^2},$$

which tends to 0 as $\lambda \to \infty$ on account of the fact that $c_\lambda \leq C/\lambda^3$, with C independent of λ.

Since continuous functions on I are uniformly continuous there, the statement about the uniform convergence follows since for these functions δ is independent of ε. By picking a sequence $\{\varepsilon_n\}$ that decreases to 0 and the corresponding $\lambda_n = N_n$ above sufficiently large integers (or $\lambda = 2^{k_n}$ to make things simpler), we get a sequence of trigonometric polynomials that converge uniformly to f on I, and (i) holds.

Next, when f is Lipschitz of order $\alpha \leq 1$, by Proposition 15, the constant $C_\alpha = C$ is independent of α, and we have

$$\left| f(x) - c_\lambda \int_I f(x - y) \, F(y, \lambda)^2 \, dy \right|$$

$$\leq c_\lambda \int_I \left| f(x) - f(x - y) \right| F(y, \lambda)^2 \, dy$$

$$\leq c_\lambda \, L \int_I |y|^\alpha F(y, \lambda)^2 \, dy \leq c \, L \, I_{\alpha, \lambda} \leq C L \, \lambda^{-\alpha},$$

which, when $\lambda = m$ and $t_m(x) = c_m \int_I f(y) F(x - y, \lambda)^2 dy$, becomes

$$|f(x) - t_m(x)| \leq CL\, m^{-\alpha}, \quad x \in I,$$

where t_m is a trigonometric polynomial of degree $2(m - 1)$. Then, given an arbitrary integer n, we take m above to be equal to $(n/2) + 1$ when n is even and $(n + 1)/2$ when n is odd, and then the polynomial defined above is of order $\leq n$ and the right-hand side $O(1/n^\alpha)$. This proves (ii).

To verify (iii), given n, let $\mathcal{P}_n$ be the partition of $[-\pi, \pi]$ in Π_e consisting of n equal length subintervals, and let g be the continuous piecewise linear function of period 2π that assumes the same values as f at the endpoints of the intervals in $\mathcal{P}_n$ and is linear between these points. The graph of g is a broken line, no segment of which has a slope greater than $\omega_f(2\pi/n)/(2\pi/n)$ in absolute value, and, consequently, $g(x)$ satisfies the assumptions of (ii) with $L = \omega_f(2\pi/n)/(2\pi/n)$. Hence, given an integer n, there is a trigonometric polynomial $t_n(x)$ of order n such that

$$|g(x) - t_n(x)| \leq C(n/2\pi)\omega_f(2\pi/n)\frac{1}{n} = (C/2\pi)\,\omega_f(2\pi/n).$$

Now, if $x \in I$, x differs by less than $2\pi/n$ from one of the numbers in I for which g is defined to be equal to f; let t_x be such a number. Then $|f(x) - g(x)| \leq |f(x) - f(t_x)| + |f(t_x) - g(t_x)| + |g(t_x) - g(x)| \leq 2\omega_f(2\pi/n)$ for all x, and, consequently,

$$|f(x) - t_n(x)| \leq |f(x) - g(x)| + |g(x) - t_n(x)| \leq ((c/2\pi) + 2)\,\omega_f(2\pi/n),$$

which is the desired estimate. Moreover, since f is uniformly continuous on I, $\lim_n \omega_f(2\pi/n) = 0$, and a refined version of the Weierstrass theorem for approximation by trigonometric polynomials has been established. Thus the proof is finished.   □

Chapter 4
The Modified Π-Riemann Sums

In this chapter, we introduce the modified Π-Riemann sums, which allow for the consideration of sets other than intervals in the partitions under consideration. Some of the applications covered include ψ-asymptotically distributed sequences, uniformly distributed sequences, and extensions of recent results on deleting items and disturbing mesh in the Riemann integral [61].

We will introduce now the modified Riemann sums [100] in the context of admissible partitions. Let Π be an admissible family of partitions of I, and let Φ be a set mapping defined on the subintervals J of I that belong to some partition in Π. We assume that Φ assigns to each such J a subset $J^1 = \Phi(J) \subset I$ with the following two properties:

(i) $\chi_{\Phi(J)}$ is Riemann integrable, and so, the length $|J^1|$ of J^1 is well-defined as $|J^1| = 0$ if $J^1 = \emptyset$, and $|J^1| = |\Phi(J)| = \int_I \chi_{\Phi(J)}$, otherwise.
(ii) There exists $\eta > 0$ such that $J^1 = \Phi(J) \subset J$, whenever $|J| < \eta$.

An observation before we proceed. Recall that $E \subseteq I$ is Jordan measurable with Jordan measure $m(E)$ iff χ_E is Riemann integrable and $m(E) = \int_I \chi_E$. Thus, (i) could be restated as requiring that $\Phi(J)$ be Jordan measurable and setting $|\Phi(J)| = m(\Phi(J))$.

Given a partition $\mathcal{P} = \{I_1, \ldots, I_m\}$ of I in Π, we will denote with $\mathcal{P}^1$ the collection of subsets of I consisting of $I_1^1 = \Phi(I_1), \ldots, I_m^1 = \Phi(I_m)$, say, and set

$$u_\Pi(f, \mathcal{P}^1) = \sum_{k=1}^m \left(\sup_{I_k^1} f \right) |I_k^1|, \quad \text{and} \quad l_\Pi(f, \mathcal{P}^1) = \sum_{k=1}^m \left(\inf_{I_k^1} f \right) |I_k^1|.$$

© The Author(s), under exclusive license to Springer Nature Switzerland AG 2022
A. Torchinsky, *A Modern View of the Riemann Integral*,
Lecture Notes in Mathematics 2309, https://doi.org/10.1007/978-3-031-11799-2_4

We call these expressions the *modified upper and lower Π-Riemann sums of f on I along P*, respectively, and set

$$u_\Pi(f) = \inf_{P \in \Pi} u_\Pi(f, P^1), \quad \text{and} \quad l_\Pi(f) = \sup_{P \in \Pi} l_\Pi(f, P^1).$$

We then say that the *modified Π-Riemann sums of f on I converge* if

$$u_\Pi(f) = l_\Pi(f). \tag{48}$$

Clearly in this case, the *(arbitrary) modified Π-Riemann sums of f on I* given by

$$s_\Pi(f, P^1, C^1) = \sum_{k=1}^{m} f(c_k^1) |I_k^1|, \quad c_k^1 \in I_k^1, \quad P \in \Pi, \tag{49}$$

which lie between $l_\Pi(f, P^1)$ and $u_\Pi(f, P^1)$, will also converge to the common limit above in a sense that will be made precise in Theorem 14.

Also, since (48) holds for the constant function χ_I, with $P = \{I_1, \dots, I_m\}$, it follows that

$$\inf_{P \in \Pi} \sum_{k=1}^{m} |I_k^1| = \sup_{P \in \Pi} \sum_{k=1}^{m} |I_k^1|. \tag{50}$$

Our next result describes the behaviour of the modified Π-Riemann sums of a Π-Riemann integrable function f on I.

Theorem 14 *Let Φ be a set mapping that satisfies (i), (ii), and (50) above, and let f be a Π-Riemann integrable function on I. Then, the modified Π-Riemann sums of f on I converge. Moreover, there is a sequence $\{P_n\}$ of partitions of I in Π such that*

$$\lim_n \left(u_\Pi(f, P_n^1) - l_\Pi(f, P_n^1) \right) = 0, \tag{51}$$

and, for any sequence $\{P_n\}$ of partitions of I in Π that satisfies (5), we have

$$u_\Pi(f) = \lim_n u_\Pi(f, P_n^1) = \lim_n l_\Pi(f, P_n^1) = l_\Pi(f).$$

Furthermore, the arbitrary modified Π-Riemann sums $s_\Pi(f, P_n^1, C)$ of f on I, defined by (49) above, also converge to $u_\Pi(f) = l_\Pi(f)$.

Proof On account of (5), given $\varepsilon > 0$, pick a partition P of I in Π consisting of intervals $I_1, \dots, I_m$, such that $|I_k| \leq |I|/n \leq \eta$ for $1 \leq k \leq m$, and

$$U_\Pi(f, P) - L_\Pi(f, P) \leq \varepsilon.$$

Then, since $|I_k| \leq \eta$, by (ii) above, $I_k^1 = \Phi(I_k) \subset I_k$ for all k, and it follows that

$$\left(\sup_{I_k^1} f\right)|I_k^1| - \left(\inf_{I_k^1} f\right)|I_k^1| \leq \left(\sup_{I_k} f - \inf_{I_k} f\right)|I_k|, \quad 1 \leq k \leq m.$$

Thus summing over k, we have

$$u_\Pi(f, \mathcal{P}^1) - l_\Pi(f, \mathcal{P}^1) \leq U_\Pi(f, \mathcal{P}) - L_\Pi(f, \mathcal{P}) \leq \varepsilon, \tag{52}$$

and since $u_\Pi(f) \leq u_\Pi(f, \mathcal{P}^1)$ and $l_\Pi(f, \mathcal{P}^1) \leq l_\Pi(f)$, (52) yields

$$u_\Pi(f) - l_\Pi(f) \leq u_\Pi(f, \mathcal{P}^1) - l_\Pi(f, \mathcal{P}^1) \leq \varepsilon,$$

which, since ε is arbitrary, gives the convergence of the modified Π-Riemann sums of f on I.

Next, since f is Π-Riemann integrable on I, there is a sequence $\{\mathcal{P}_n\}$ of partitions of I in Π that satisfies (5). Pick N so that $|I|/N \leq \eta$, and note that for each $I_\ell \in \mathcal{P}_n$ with $n \geq N$, we have $|I_\ell| \leq \eta$ and so $\Phi(I_\ell) \subset I_\ell$, and, consequently, as in (52) above, it follows that

$$u_\Pi(f, \mathcal{P}_n^1) - l_\Pi(f, \mathcal{P}_n^1) \leq U_\Pi(f, \mathcal{P}_n) - L_\Pi(f, \mathcal{P}_n), \quad \text{all } n \geq N, \tag{53}$$

and since the right-hand side of (53) tends to 0, so does the left-hand side, and (51) holds.

Furthermore, since $l_\Pi(f, \mathcal{P}_n^1) \leq l_\Pi(f) \leq u_\Pi(f) \leq u_\Pi(f, \mathcal{P}_n^1)$ for all n, it follows that

$$0 \leq \max\{u_\Pi(f, \mathcal{P}_n^1) - u_\Pi(f), \, l_\Pi(f) - l_\Pi(f, \mathcal{P}_n^1)\} \leq u_\Pi(f, \mathcal{P}_n^1) - l_\Pi(f, \mathcal{P}_n^1),$$

and since the right-hand side above tends to 0, so does the left-hand side, and therefore,

$$\lim_n u_\Pi(f, \mathcal{P}_n^1) = u_\Pi(f), \quad \text{and,} \quad \lim_n l_\Pi(f, \mathcal{P}_n^1) = l_\Pi(f).$$

Also,

$$\lim_n l_\Pi(f, \mathcal{P}_n^1) = \lim_n u_\Pi(f, \mathcal{P}_n^1) - \lim_n \left(u_\Pi(f, \mathcal{P}_n^1) - l_\Pi(f, \mathcal{P}_n^1)\right) = u_\Pi(f) = l_\Pi(f).$$

Finally, since for an arbitrary modified Riemann sum $s_\Pi(f, \mathcal{P}_n^1, C^1)$ of f, we have $l_\Pi(f, \mathcal{P}_n^1) \leq s_\Pi(f, \mathcal{P}_n^1) \leq u_\Pi(f, \mathcal{P}_n^1)$, it follows that

$$\lim_n s_\Pi(f, \mathcal{P}_n^1, C^1) = u_\Pi(f), \tag{54}$$

and we have finished. $\qquad\qquad\qquad\qquad\qquad\qquad\qquad\qquad\qquad\qquad\qquad\qquad\qquad\square$

In the applications that follow, we will often consider the admissible family Π to be Π_e, the family of all partitions of I consisting of n intervals of equal length, all n.

We examine first an instance where $|\Phi(J)|$ is a function of $|J|$, linear for the example that motivated our results [64], but not necessarily in general.

We then have:

Example 1 With $\{\beta_n\}$ a sequence of nonnegative numbers such that $0 \leq \beta_n \leq 1/n$, for all n, let the mapping Φ be defined on the subintervals of Π_e as follows: If $\mathcal{P}_n = \{I_k^n\}$, then for $1 \leq k \leq n$, $\Phi(I_k^n) = I_k^{n,1} \subset I_k^n$, where $\chi_{\Phi(I_k^{n,1})}$ is Riemann integrable, and $|I_k^{n,1}| = \beta_n$.

Note that for the constant function χ_I, with $\mathcal{P} = \{I_k^n\}$, as in (50), we have

$$u_{\Pi_e}(\chi_I) = \inf_{\mathcal{P} \in \Pi_e} \sum_{k=1}^n \beta_n = \inf_{\mathcal{P} \in \Pi_e} n\beta_n$$

$$= \sup_{\mathcal{P} \in \Pi_e} n\beta_n = \sup_{\mathcal{P} \in \Pi_e} \sum_{k=1}^n \beta_n = l_{\Pi_e}(\chi_I),$$

and, therefore, we assume that $\lim_n n\beta_n = \eta$ exists.

Then it follows that

$$s_{\Pi_e}(f, \mathcal{P}_n^1, C^1) = \sum_{k=1}^n f(c_k^1) |\Phi(I_k^n)|$$

$$= \sum_{k=1}^n f(c_k^1) \frac{|\Phi(I_k^n)|}{|I_k^n|} |I_k^n| = n\beta_n \, S_{\Pi_e}(f, \mathcal{P}_n, C^1),$$

where $\lim_n n\beta_n = \eta$, and $\lim_n S_{\Pi_e}(f, \mathcal{P}_n, C^1) = \int_I f$. Thus, if f is Π_e-Riemann integrable on I, by (54), it follows that

$$u_{\Pi_e}(f) = \lim_n s_{\Pi_e}(f, \mathcal{P}_n^1, C^1) = \eta \int_I f.$$

We may think of the β_n as the image of a nonnegative function ϕ defined in a neighbourhood of the origin so that on the set $\{1/n : n \in \mathbb{N}\}$ satisfies $n\,\phi(1/n) \leq 1$, $\phi(0) = 0$, and such that $\lim_n n\,\phi(1/n) = \eta$, exists. Then $\lim_n n\,\phi(1/n)$ corresponds to $\phi'_+(0)$, the right-hand derivative of ϕ at the origin [100]. The choice $\phi(t) = t/2$ corresponds to the instance discussed in [64].

We will next discuss an instance where the mapping Φ depends on the location of the intervals and not on their measure.

Example 2 Let $\{\alpha_{k,n}\}$ be a double sequence of nonnegative real numbers defined for $1 \leq k \leq n$ and $1 \leq n < \infty$, such that $0 \leq \alpha_{k,n} \leq 1$ for all k, and let the mapping Φ be defined on the subintervals of the family $\Pi_e = \{\mathcal{P}_n\}$ as follows: If $\mathcal{P}_n = \{I_k^n\}$, then for $1 \leq k \leq n$, $\Phi(I_k^n) = I_k^{n,1} \subset I_k^n$, where $\chi_{I_k^{n,1}}$ is Riemann integrable, and $|I_k^{n,1}| = \alpha_{k,n} |I_k^n|$ if $\alpha_{k,n} \neq 0$, and $\Phi(I_k^n) = \emptyset$ if $\alpha_{k,n} = 0$.

First, as in (50), it follows that

$$\inf_{\mathcal{P} \in \Pi_e} \frac{1}{n} \sum_{k=1}^n \alpha_{k,n} = \sup_{\mathcal{P} \in \Pi_e} \frac{1}{n} \sum_{k=1}^n \alpha_{k,n},$$

and therefore, the limit

$$\lim_n \frac{1}{n} \sum_{k=1}^n \alpha_{k,n} = \alpha \tag{55}$$

exists. It becomes quickly apparent that (55) does not suffice in this case, so we additionally assume that for some $\eta > 0$, provided that $m/n \geq \eta > 0$,

$$\lim_{m,n \to \infty} \frac{1}{m} \sum_{k=1}^m \alpha_{k,n} = \alpha. \tag{56}$$

We then have:

Proposition 16 *Let I be a finite closed interval of $\mathbb{R}$, and let the double sequence $\{\alpha_{k,n}\}$ satisfy (56). Then, if Φ is defined as in Example 2, for a Π_e-Riemann integrable function f on I, we have*

$$u_{\Pi_e}(f) = \alpha \int_I f. \tag{57}$$

Proof We begin by proving that if for partitions $\mathcal{P}_n = \{I_k^n\}$ of I in Π_e, we let

$$\psi_n(x) = \sum_{k=1}^n \alpha_{k,n} \chi_{I_k^n}(x), \tag{58}$$

then, for an interval $J \subset I$,

$$\lim_n \int_J \psi_n = \lim_n \int_I \chi_J \psi_n = \alpha |J|. \tag{59}$$

As in the proof of Proposition 12, it suffices to verify (59) for $J = [a, d] \subset I = [a, b]$, with $d \leq b$. Fix n, and pick $m < n$ such that

$$m/n \leq (d - a) < (m + 1)/n. \tag{60}$$

Note that (60) implies that

$$m/n \geq (d - a) - 1/n,$$

and, in particular, $m \to \infty$ as $n \to \infty$. Moreover, since for n large enough $(d - a) - 1/n \geq (d - a)/2$, we pick $\eta = (d - a)/2$ and restrict ourselves to m, n sufficiently large so that $m/n \geq \eta$.

Then, by (58) and (60), we have

$$\int_J \psi_n = \sum_{k=1}^n \alpha_{k,n} \int_{J \cap I_k} dx$$

$$= \sum_{k=1}^m \alpha_{k,n} \int_{J \cap I_k} dx + \sum_{k=m+1}^n \alpha_{k,n} \int_{J \cap I_k} dx$$

$$= s_1(n) + s_2(n),$$

say. Since $(d - a) < (m + 1)/n$, at most one summand in $s_2(n)$ is not 0, and $s_2(n) = 0(1/n)$. As for $s_1(n)$, by (60), we have

$$(d - a) \frac{m}{m + 1} \frac{1}{m} \sum_{k=1}^m \alpha_{k,n} = (d - a) \frac{1}{m + 1} \sum_{k=1}^m \alpha_{k,n}$$

$$< s_1(n) = \frac{1}{n} \sum_{k=1}^m \alpha_{k,n} \leq (d - a) \frac{1}{m} \sum_{k=1}^m \alpha_{k,n},$$

and, therefore, taking the limit as $n, m \to \infty$, by (60), it follows that

$$(d - a) \alpha \leq \lim_n s_1(n) \leq (d - a) \alpha,$$

and, therefore, (59) holds in this case. Then, as in the proof of Proposition 12, the same is true for an arbitrary interval $J = [c, d] \subset I$.

Next, observe that with ψ_n as in (58), $\{\psi_n\}$ is a uniformly bounded sequence of Riemann integrable functions on I, and for a Riemann integrable function f, and

$\mathcal{P}_n = \{I_k^n\}$ in Π_e,

$$\int_I f \psi_n = \sum_{k=1}^{n} \alpha_{k,n} \int_{I_k^n} f = \sum_{k=1}^{n} \alpha_{k,n} \int_{I_k^n} f \mp f(c_k^{n,1}) |I_k|$$

$$= \sum_{k=1}^{n} \alpha_{k,n} \int_{I_k^n} (f - f(c_k^{n,1})) + \sum_{k=1}^{n} \alpha_{k,n} f(c_k^{n,1}) |I_k| = A_1(n) + A_2(n),$$

say. Now, since

$$|A_1(n)| \leq \sum_{k=1}^{n} \alpha_{k,n} \, \mathrm{osc}\,(f, I_k^n) \, |I_k^n| \leq \sum_{k=1}^{n} \mathrm{osc}\,(f, I_k^n) \, |I_k^n|,$$

by (6), $\lim_n A_1(n) = 0$.

Moreover, since $A_2(n) = s_{\Pi_e}(f, \mathcal{P}_n^1, C^1)$, by (54), it readily follows that

$$u_{\Pi_e}(f) = \lim_n s_{\Pi_e}(f, \mathcal{P}_n^1, C^1)$$

$$= \lim_n A_2(n) = \lim_n \int_I f \psi_n = \alpha \int_I f,$$

(57) holds, and the proof is finished.   □

The results concerning the deleting item theorem and disturbing mesh theorems in [61] follow from Proposition 16. A word about the latter, which is proved for nonnegative functions in [61]. With the notation in that paper, let $\{d_{n,k}\}$ be a double sequence of nonnegative real numbers, and for partitions $\mathcal{P}_n = \{I_1^n, \dots, I_n^n\}$ of I in Π_e, let $\Phi(I_k^n)$ be defined as follows: Put $\Phi(I_k^n) = I_k^{n,1} \subset I_k^n$, with $\chi_{\Phi(I_k^n)}$ Riemann integrable, and

$$|I_k^{n,1}| = \frac{|I|}{n + d_{n,k}} = \frac{n}{n + d_{n,k}} |I_k^n|, \quad 1 \leq k \leq n, n = 1, 2, \dots$$

Then, by Proposition 16 with $\alpha_{k,n} = n/(n + d_{n,k})$ there, provided that

$$\lim_{m,n \to \infty} \frac{1}{m} \sum_{k=1}^{m} \frac{n}{n + d_{n,k}} = \alpha \leq 1,$$

it follows that $u_{\Pi_e}(f) = \lim_n s_{\Pi_e}(f, \mathcal{P}_n^1, C^1) = \alpha \int_I f$. For instance, if $d_{n,k} = O(n^\rho)$ for $0 < \rho < 1$, this holds with $\alpha = 1$.

As for the deleting item theorem, we consider the following setting. Let K and L be fixed integers independent of n, and let $\alpha_{k,n} = 0$ for $1 \le k \le K$ and $n - L \le k \le n$, and $\alpha_{k,n} = 1$ otherwise. Then (56) holds with $\alpha = 1$ there, and consequently, $u_{\Pi_e}(f) = \int_I f$. It is clear that variants of this result hold by considering different subsets $J_K \subset \mathbb{N}$ of indices where the $\alpha_{k,n} = 0$.

Furthermore, the idea of the proof can be adapted to the following setting, with its many interesting applications [89]. Let $I = [a, b]$ be a bounded interval in $\mathbb{R}$, and let $\{x_k^n\}$ be a collection of points in I such that $a = x_0^n < x_1^n < \ldots < x_n^n = b$, $1 \le k \le n$, all n. Let $J_k^n = [x_{k-1}^n, x_k^n]$, $1 \le k \le n$, and consider the partitions $\mathcal{P}_n = \{J_k^n\}$ of I, where we assume that $\lim_n \|\mathcal{P}_n\| = 0$. Suppose further that ψ is Riemann integrable on I, and let K and L be fixed integers independent of n. We say that $\{x_k^n\}$ is ψ-asymptotically equidistributed on I, if for all k such that $0 < K < k < n - L$,

$$\psi(x_k^n)(x_{k+1}^n - x_k^n) = \frac{|I|}{n} + \frac{\varepsilon_{n,k}}{n}, \qquad |\varepsilon_{n,k}| \le \varepsilon_n, \tag{61}$$

with ε_n independent of k, and $\lim_n \varepsilon_n = 0$.

Now, by a simple variant of the deleting item theorem, from (61), it follows that $\int_I \psi = |I|$, and so we assume that

$$\frac{1}{|I|} \int_I \psi = 1, \tag{62}$$

which also follows by setting $f = \chi_I$ in (63) below.

We then have:

Theorem 15 *Let $\{x_k^n\}$ be ψ-asymptotically equidistributed on I, where ψ satisfies (58). Then, for a Riemann integrable f on I, we have*

$$\lim_n \frac{1}{n} \sum_{k=1}^n f(x_k^n) = \frac{1}{|I|} \int_I f \, \psi. \tag{63}$$

Proof From (61), we get

$$\frac{1}{n} \sum_{k=1}^n f(x_k^n) = \frac{1}{n} \sum_{k=1}^K f(x_k^n) + \frac{1}{|I|} \sum_{k=K+1}^{n-(L-1)} f(x_k^n)\psi(x_k^n) \, |J_k^n|$$

$$- \frac{1}{n} \frac{1}{|I|} \sum_{k=K+1}^{n-(L-1)} \varepsilon_{k,n} + \frac{1}{n} \sum_{k=n-L}^n f(x_k^n),$$

and, therefore, with M_f a bound for f,

$$\left| \frac{1}{n} \sum_{k=1}^{n} f(x_k^n) - \frac{1}{|I|} \sum_{k=K+1}^{n-L} f(x_k^n) \psi(x_k^n) |J_k^n| \right|$$

$$\leq \frac{1}{n} M_f K + \frac{1}{|I|} \frac{1}{n} \varepsilon_n (n - (K + L)) + \frac{1}{n} M_f L, \tag{64}$$

where the right-hand side goes to 0 as $n \to \infty$.

Now, the expression in the sum is essentially that in the deleting item theorem and, with M_ψ, a bound for ψ, is bounded accordingly,

$$\left| \frac{1}{|I|} \sum_{k=1}^{n} f(x_k^n) \psi(x_k^n) |J_k^n| - \frac{1}{|I|} \sum_{k=K+1}^{n-L} f(x_k^n) \psi(x_k^n) |J_k^n| \right|$$

$$\leq \frac{1}{|I|} M_f M_\psi \|\mathcal{P}_n\| K + \frac{1}{|I|} M_f M_\psi \|\mathcal{P}_n\| L, \tag{65}$$

where, since $\lim_n \|\mathcal{P}_n\| = 0$, the right-hand side goes to 0 as $n \to \infty$.

Whence, combining (64) and (65), we have

$$\limsup_n \left| \frac{1}{n} \sum_{k=1}^{n} f(x_k^n) - \frac{1}{|I|} \sum_{k=1}^{n} f(x_k^n) \psi(x_k^n) |J_k^n| \right| = 0,$$

and, consequently, since the subtrahend in the above expression converges to $(1/|I|) \int_I f\psi$, it follows that

$$\lim_n \frac{1}{n} \sum_{k=1}^{n} f(x_k^n) = \frac{1}{|I|} \lim_n S_\Pi(f\psi, \mathcal{P}_n, C) = \frac{1}{|I|} \int_I f \psi,$$

(63) holds, and we have finished.   □

We consider next the case when the double sequence $\alpha_{k,n} = \alpha_k$ is independent of the level n, which includes the extension of the patterns alluded to above. Specifically, we have:

Example 3 Let $\{\alpha_k\}$ be a sequence of nonnegative real numbers such that $0 \leq \alpha_k \leq 1$ for all k, and let the mapping Φ be defined on the subintervals of the family $\Pi_e = \{\mathcal{P}_n\}$ as follows: If $\mathcal{P}_n = \{I_k^n\}$, then $\Phi(I_k^n) = I_k^{n,1} \subset I_k^n$, where $\chi_{I_k^{n,1}}$ is Riemann integrable, and $|I_k^{n,1}| = \alpha_k |I_k^n|$ if $\alpha_k \neq 0$, and $\Phi(I_k^n) = \emptyset$ if $\alpha_k = 0$, for $1 \leq k \leq n, n = 1, 2, \ldots$

In this case, the assumption is that

$$\lim_n \frac{1}{n} \sum_{k=1}^{n} \alpha_k = \alpha \tag{66}$$

exists. Then, the analogous result to Proposition 11 holds with similar, yet simpler, proof, and by (54), we have

$$u_{\Pi_e}(f) = \lim_n s_{\Pi_e}(f, \mathcal{P}_n^1, C^1) = \lim_n \sum_{k=1}^{n} f(c_k^{n,1}) |I_k^{n,1}|$$

$$= \lim_n \sum_{k=1}^{n} \alpha_k f(c_k^{n,1}) |I_k^n| = \alpha \int_I f. \tag{67}$$

4.1   Uniformly Distributed Sequences

We will now consider the notion of the distribution of a sequence. Let $I = [0, 1]$. We may think of a real sequence $\{x_k\}$ as being contained in I by replacing when necessary x_k by its fractional part $x_k - [x_k]$, where $[x_k]$ denotes the greatest integer less than or equal to x_k. We then say that a sequence $\{x_k\} \subset I$ is *uniformly distributed mod* 1 (often abbreviated u.d. mod 1, or plainly u.d.) if for all $0 \leq a \leq b \leq 1$,

$$\lim_n \frac{1}{n} \sum_{k=1}^{n} \chi_{[a,b]}(x_k) = b - a. \tag{68}$$

Informally, $\{x_k\}$ is u.d. if for any subinterval J of I, the frequency with which the fractional parts of the x_k lie in J is equal to $|J|$. This in particular implies that a u.d. sequence $\{x_k\}$ is dense in I, for if not there are $x \in I$ and an open interval $J \subset I$ containing x that contains only finitely many of the x_k, thus violating the frequency property.

U.d. sequences are characterized as follows, see [56] and Theorem 34 below: $\{x_k\}$ is uniformly distributed on I iff for all Riemann integrable functions φ defined on I,

$$\lim_n \frac{1}{n} \sum_{k=1}^{n} \varphi(x_k) = \int_I \varphi. \tag{69}$$

In this case, the Riemann integral is more adequate than the Lebesgue integral since (69) does not necessarily hold for Lebesgue integrable functions. Take, e.g., $\varphi = \chi_S$, where $S = \{x_k\}$; then, the left-hand side of (69) is equal to 1, whereas, the right-hand side, since S is countable and the Lebesgue integral of φ is then 0, is equal to 0.

Now, by linearity, (68) holds for the linear span of the characteristic functions of the subintervals of I, which contains the step functions on I and approximates the Riemann integrable functions there. Thus, it is expected that (69) will be true. Since also continuous functions approximate integrable functions on I, by the Weierstrass theorems, the following result is natural:

Theorem 16 *The following are equivalent:*

(i) *The sequence $\{x_k\}$ is uniformly distributed mod 1.*
(ii) *If f is a continuous function on I,*

$$\lim_N \frac{1}{N} \sum_{k=1}^{N} f(x_k) = \int_I f.$$

(iii) *For all $n = 0, 1, 2, \ldots$,*

$$\lim_N \frac{1}{N} \sum_{k=1}^{N} x_k^n = \frac{1}{n+1}.$$

Proof (i) implies (ii). It follows from (68) along the lines of Theorem 34.

(ii) implies (iii). It follows by applying (ii) to the continuous functions $f(x) = x^n$, $n = 0, 1, 2, \ldots$

(iii) implies (ii). By linearity, it readily follows that

$$\lim_N \sum_{k=1}^{N} g(x_k) = \int_I g,$$

for all algebraic polynomials g on I.

We claim the same is true for continuous functions f defined on I. Given $\varepsilon > 0$, by the Weierstrass theorem, pick an algebraic polynomial g such that $|f(x) - g(x)| \le \varepsilon$ throughout I, and so that $\int_I |f - g| \le \varepsilon$ as well. It then follows that

$$-\varepsilon + \frac{1}{n} \sum_{k=1}^{n} g(x_k) \le \frac{1}{n} \sum_{k=1}^{n} f(x_k) \le \frac{1}{n} \sum_{k=1}^{n} g(x_k) + \varepsilon,$$

and, therefore, since g is an algebraic polynomial,

$$-\varepsilon + \int_I g \le \liminf_n \frac{1}{n} \sum_{k=1}^{n} f(x_k) \le \limsup_n \frac{1}{n} \sum_{k=1}^{n} f(x_k) \le \int_I g + \varepsilon.$$

Hence,

$$\limsup_n \frac{1}{n} \sum_{k=1}^n f(x_k) - \liminf_n \frac{1}{n} \sum_{k=1}^n f(x_k) \le 2\varepsilon,$$

and the limit exists.

Moreover, since also $\int_I f - \varepsilon \le \int_I g \le \int_I f + \varepsilon$, it follows that

$$\int_I f - \varepsilon \le \int_I g \le \lim_n \frac{1}{n} \sum_{k=1}^n f(x_k) + \varepsilon \le \int_I g + 2\varepsilon \le \int_I f + 3\varepsilon,$$

and

$$\left| \int_I f - \lim_n \frac{1}{n} \sum_{k=1}^n f(x_k) \right| \le 3\varepsilon,$$

which gives (ii).

(ii) implies (i). Consider an interval $[a, b] \subset [0, 1]$. Since $[a, b] = [0, b] \setminus [0, a)$, and since the proof for $[0, b]$ and $[0, b)$ is identical, it suffices to consider $[0, b]$ with $b < 1$, since (68) is trivial when $b = 1$. We claim that, given $0 < \varepsilon \le \min\big(b, (1 - b)/2\big)$, we can find continuous functions g, G on I such that

$$g(x) \le \chi_{[0,b]}(x) \le G(x), \ x \in I, \quad \text{and,} \quad \int_I (G - g) \le \varepsilon.$$

This fact can be verified graphically, or otherwise, as follows. Given $0 < \varepsilon < 1 - b$, define $g(x) = 1$ if $0 \le x \le b - \varepsilon$, linear between $(b - \varepsilon, 1)$ and $(b, 0)$, and $= 0$ for $x \in [b, 1]$, and let $G(x)$ be defined to be $= 1$ for $0 \le x \le b$, linear between $(b, 1)$ and $(b + \varepsilon, 0)$, and $= 0$ for $x \in [b + \varepsilon, 1]$. These functions satisfy the conditions listed above.

We then have

$$\int_I g = \liminf_n \frac{1}{n} \sum_{k=1}^n g(x_k) \le \liminf_n \frac{1}{n} \sum_{k=1}^n \chi_{[0,b]}(x_k)$$

$$\le \limsup_n \frac{1}{n} \sum_{k=1}^n \chi_{[0,b]}(x_k) \le \limsup_n \frac{1}{n} \sum_{k=1}^n G(x_k) = \int_I G,$$

and so

$$\limsup_n \frac{1}{n} \sum_{k=1}^n \chi_{[0,b]}(x_k) - \liminf_n \frac{1}{n} \sum_{k=1}^n \chi_{[0,b]}(x_k) \le \int_I G - \int_I g \le \varepsilon,$$

which implies that the limit exists and that

$$\int_I g \le \lim_n \frac{1}{n} \sum_{k=1}^{n} \chi_{[0,b]}(x_k) \le \int_I G.$$

Combining this observation with the fact that $\int_I g \le \int_I \chi_{[0,b]} \le \int_I G$, we get

$$\left| \lim_n \frac{1}{n} \sum_{k=1}^{n} \chi_{[0,b]}(x_k) - \int_I \chi_{[0,b]} \right| \le \varepsilon,$$

which proves the claim. Thus (68) holds, and the proof is finished.   □

We also have:

Weyl's Criterion *The following are equivalent:*

(i) *The sequence $\{x_k\}$ is uniformly distributed mod 1.*
(ii) *If f is a continuous function on I with $f(0) = f(1)$, we have*

$$\lim_n \frac{1}{n} \sum_{k=1}^{n} f(x_k) = \int_I f.$$

(iii) *For each nonzero integer h, we have*

$$\lim_N \frac{1}{N} \sum_{k=1}^{N} \cos(2\pi h x_k) = \lim_N \frac{1}{N} \sum_{k=1}^{N} \sin(2\pi h x_k) = 0.$$

Proof (i) implies (ii). Follows from (69) for the particular class of functions described.

(ii) implies (iii). Apply (ii) to $f(x) = \cos(2\pi h x_k)$, $\sin(2\pi h x_k)$, which is verified since $\int_I \cos(2\pi h x)\, dx = \int_I \sin(2\pi h x)\, dx = 0$ for all integers $h \ne 0$.

(iii) implies (ii). Assume first that $\int_I f = 0$. In this case, the polynomials guaranteed by the Weierstrass theorem may be assumed to be of the form $p_n(x) = \sum_{k=1}^{n} a_k \cos(2\pi k x) + b_k \sin(2\pi k x)$. Now, if (iii) holds, then

$$\lim_N \frac{1}{N} \sum_{k=1}^{N} g(x_k) = 0,$$

for any such trigonometric polynomials. Then, given $\varepsilon > 0$, there is a trigonometric polynomial g such that $|f(x) - g(x)| \le \varepsilon/3$ throughout I and $\int_I |f - g| \le \varepsilon/3$.

Now,

$$\left| \int_I f - \frac{1}{N} \sum_{k=1}^{N} f(x_k) \right| \leq \int_I |f - g| + \left| \int_I g - \frac{1}{N} \sum_{k=1}^{N} g(x_k) \right|$$

$$+ \left| \frac{1}{N} \sum_{k=1}^{N} g(x_k) - \frac{1}{N} \sum_{k=1}^{N} f(x_k) \right|,$$

where the first and third terms on the right-hand side are each less than $\varepsilon/3$ whatever the value of N because of the choice of g. Since $\int_I g = 0$, by taking N sufficiently large, the second term on the right-hand side is less than $\varepsilon/3$ by the assumption (iii). Thus, the whole sum does not exceed ε, and the proof is complete in this case. To complete the proof in the general case, observe that the conclusion holds trivially for the function χ_I and apply the result we just proved to the function $f - (\int_I f) \chi_I$, which has integral 0.

(ii) implies (i). It suffices to prove (68) for $\chi_{[0,b]}$ where $0 < b < 1$. We claim that, given $\varepsilon > 0$, we can find continuous functions g, G (with $g(0) = g(1)$, $G(0) = G(1)$) such that

$$g(x) \leq \chi_{[0,b]}(x) \leq G(x), \ x \in I, \quad \text{and,} \quad \int_I (G - g) \leq \varepsilon.$$

This fact can be verified graphically, or otherwise as follows. For $\eta > 0$ sufficiently small, define $g(x) = 0$ if $x \in [0, \eta]$, linear between $(\eta, 0)$ and $(2\eta, 1)$, $= 1$ for $x \in (2\eta, b - \eta)$, linear between $(b - \eta, 1)$ and $(b, 0)$, and $= 0$ for $x \in [b, 1]$. As for $G(x)$, it is $= 1$ for $0 \leq x \leq b$, linear between $(b, 1)$ and $(b + \eta, 0)$, $= 0$ for $x \in (b + \eta, 1 - \eta)$, and linear between $(1 - \eta, 0)$ and $(1, 1)$. Hence, $\int_I (G - g) = 4\eta \leq \varepsilon$ provided that $\eta = \varepsilon/4$. The proof is now completed exactly as the proof of (ii) implies (i) of the previous theorem. Thus (68) holds and the sequence is u.d. mod 1. □

We consider the sequence $\{k\alpha\}$, $k = 1, 2, \ldots$, where α is irrational. Then for integers $h \neq 0$, by the Lagrange identities, we have

$$\frac{1}{N} \sum_{k=1}^{N} \cos(2\pi h\alpha k) = \frac{1}{N} \left(-\frac{1}{2 \sin(\pi h\alpha)} \left(\sin(\pi h\alpha) - \sin\left((2N + 1)\pi h\alpha \right) \right) \right),$$

and, consequently, since $\sin(\pi h\alpha) \neq 0$,

$$\limsup_N \left| \frac{1}{N} \sum_{k=1}^{N} \cos(2\pi h\alpha k) \right| \leq \limsup_N \frac{1}{N} \frac{1}{|\sin(\pi h\alpha)|} = 0.$$

Similarly, since

$$\sum_{k=1}^{N} \sin(2\pi h\alpha k) = \frac{1}{2\sin(\pi h\alpha)}\left(\cos(\pi h\alpha) - \cos\left((2N+1)\pi\alpha\right)\right),$$

$$\limsup_{N}\left|\frac{1}{N}\sum_{k=1}^{N}\sin(2\pi h\alpha k)\right| \leq \limsup_{N}\frac{1}{N}\frac{1}{|\sin(\pi h\alpha)|} = 0.$$

Thus (ii) of the Weyl's criterion holds, and $\{k\alpha\}$ is u.d. mod 1.

Other interesting instances of u.d. sequences can be obtained from (69):

Theorem 17 *Let $\{\mathcal{P}_n\}$ be a collection of partitions of I in Π_e, one for each n, and let $\{C_n\}$ be a corresponding collection of tags. Then, the points in this collection of tags can be ordered as a u.d. sequence $\{x_k\}$.*

Proof First we order the intervals of the partitions $\mathcal{P}_n$ in Π_e sequentially as follows. We say that I_h^m precedes I_k^n, provided that $m < n$, or, if $m = n$, when $h < k$. Call the resulting ordered sequence $\{I_n\}$. Also, any collection $C_n = \{c_1^n, \ldots, c_n^n\}$ of tags in $\mathcal{P}_n$ can be ordered as above so that for the ordered sequence $\{x_n\}$, we have $x_n \in I_n$, all n.

Consider then the expression

$$\frac{1}{N}\sum_{k=1}^{N}\varphi(x_k),$$

where $N = (1 + \cdots + n) + m = s_n + m, 0 \leq m < n$. A moment's thought shows that

$$\frac{1}{N}\sum_{k=1}^{N}\varphi(x_k) = \frac{1}{N}\sum_{k=1}^{n}k\,S(\varphi, \mathcal{P}_k, C_k) + \frac{1}{N}\sum_{k=s_n+1}^{N}\varphi(x_k),$$

where the second summand, with M_φ a bound for φ and since $N \geq s_n = n(n+1)/2$, satisfies

$$\frac{1}{N}\left|\sum_{k=s_n+1}^{N}\varphi(x_k)\right| \leq \frac{2}{n(n+1)}M_\varphi\left(s_n + m - (s_n+1)\right)$$

$$\leq 2M_\varphi/(n+1) \to 0 \quad \text{as } N \to \infty.$$

As for the first term, it can be written as

$$\frac{s_n}{N}\frac{1}{s_n}\sum_{k=1}^{n}k\,S(\varphi, \mathcal{P}_k, C_k).$$

Now, since for a Riemann integrable φ, $S(\varphi, \mathcal{P}_k, C_k) \to \int_I \varphi$ as $k \to \infty$, given $\varepsilon > 0$, there is n_ε such that $\left| S(\varphi, \mathcal{P}_k, C_k) - \int_I \varphi \right| \leq \varepsilon$ for $k \geq n_\varepsilon$. Also, since convergent sequences are bounded, $|S(\varphi, \mathcal{P}_k, C_k)| \leq M$ for all k, in particular for $k \leq n_\varepsilon$. Then, since $(1/s_n) \sum_{k=1}^n k = 1$, it readily follows that

$$\left| \frac{1}{s_n} \sum_{k=1}^n k S(\varphi, \mathcal{P}_k, C_k) - \int_I \varphi \right| = \left| \frac{1}{s_n} \sum_{k=1}^n k \left(S(\varphi, \mathcal{P}_k, C_k) - \int_I \varphi \right) \right|$$

$$\leq \left| \frac{1}{s_n} \sum_{k=1}^{n_\varepsilon} k \left(S(\varphi, \mathcal{P}_k, C_k) - \int_I \varphi \right) \right| + \left| \frac{1}{s_n} \sum_{k=n_\varepsilon+1}^{n} k \left(S(\varphi, \mathcal{P}_k, C_k) - \int_I \varphi \right) \right|.$$

As for the first term, it is bounded by

$$\left(M_\varphi + \left| \int_I \varphi \right| \right) \frac{1}{s_n} \sum_{k=1}^{n_\varepsilon} k,$$

which goes to 0 as N, and consequently, n tends to ∞. And the second term is bounded by $(\varepsilon/s_n) \sum_{k=1}^{n_\varepsilon} k \leq \varepsilon$ and, therefore, (69) holds. Since φ is an arbitrary Riemann integrable function, $\{x_k\}$ is u.d. mod 1.   □

If the sequence $\{x_n\}$ is dense in I, a moment's thought shows that we may rearrange it so that $x_1 \in [0, 1/2]$, $x_2 \in [1/2, 1]$, $x_3 \in [0, 1/3]$, $x_4 \in [1/3, 2/3]$, and so on. This sequence is u.d. by the argument of Theorem 17.

De Bruijn and Post observed that if the limit of the averages in (69) exists for every u.d. sequence in I, then φ is Riemann integrable on I, [14]. Salvati and Volčič described the behaviour of the averages when φ is bounded but not Riemann integrable as follows. Let Λ denote the set of all limit points of the Riemann sums $\{S(\varphi, \mathcal{P}_k, C_k)\}$; Λ will be called the set of all *Riemann integrals* of φ. Then, Λ is a closed and bounded interval with endpoints $L(\varphi)$, the lower Darboux integral of φ, and $U(\varphi)$, the upper Darboux integral of φ. Furthermore, for any $\lambda', \lambda'' \in \Lambda$ such that $\lambda' \leq \lambda''$, there exists a u.d. sequence $\{x_k\}$ such that the set of limits of all convergent subsequences of $\{(1/n) \sum_{k=1}^n \varphi(x_k)\}$ is exactly $[\lambda', \lambda'']$.

Here we will prove that there are u.d. mod 1 sequences so that the corresponding averages in (69) converge to the limsup and liminf, and that we can fill Λ by dovetailing these sequences [84].

We begin by proving:

Proposition 17 *Let φ be a bounded function defined on $I = [0, 1]$. Then there are u.d. sequences $\{x_k\}$ and $\{y_k\}$ such that*

$$L_{\Pi_e}(\varphi) = \lim_n \frac{1}{n} \sum_{k=1}^n \varphi(x_k), \quad \text{and,} \quad U_{\Pi_e}(\varphi) = \lim_n \frac{1}{n} \sum_{k=1}^n \varphi(y_k).$$

Proof Since $L_{\Pi_e}(\varphi) = -U_{\Pi_e}(-\varphi)$, it suffices to prove the first statement. Now, as in the proof of Theorem 1, let $\{\mathcal{P}_n\}$ be a sequence of partitions of I in Π_e such that for appropriate tags $\{C_n\}$, $\lim_n S_{\Pi_e}(\varphi, \mathcal{P}_n, C_n) = U_{\Pi_e}(\varphi)$. Then, by Theorem 14,

$$\lim_n \frac{1}{n} \sum_{k=1}^{N} \varphi(y_k) = U_{\Pi_e}(\varphi),$$

and the proof is finished.   □

As for the *dovetailing*, we say that the sequence $\{t_k\}$ is the dovetailing of the bounded real sequences $\{r_k\}$ and $\{s_k\}$, if $\{r_k\}$ and $\{s_k\}$ are subsequences of $\{t_k\}$ corresponding to a partition of $\mathbb{N}$ into two infinite subsets.

We will now show some relations between the averages of two sequences $\{r_k\}$ and $\{s_k\}$ and the averages of their dovetailings.

Theorem 18 *Let $\{r_k\}$ and $\{s_k\}$ be bounded real sequences such that*

$$\lim_n \frac{1}{n} \sum_{k=1}^{n} r_k = r < s = \lim_n \frac{1}{n} \sum_{k=1}^{n} s_k.$$

Then for any $t \in (r, s)$, there is a dovetailing $\{t_k\}$ of $\{r_k\}$ and $\{s_k\}$ such that

$$\lim_n \frac{1}{n} \sum_{k=1}^{n} t_k = t.$$

Moreover, any dovetailing $\{t_k\}$ of u.d. bounded sequences $\{r_k\}$, $\{s_k\}$ is also u.d.

Proof Let $0 < \lambda < 1$ be such that $t = \lambda r + (1 - \lambda)s$, and for an integer n, let

$$m(n) = \max\{m \in \mathbb{N} : m/n < \lambda\}.$$

Then $\{m(n)\}$ is an increasing sequence, which by the density of the rationals satisfies $\lim_n m(n)/n = \lambda$, and, consequently $\lim_n m(n) = \infty$. Moreover, since $n - m(n) > n(1 - \lambda)$, also $\lim_n n - m(n) = \infty$.

We will define the sequence $\{t_n\}$ inductively. Since $\lambda < 1$, we have $m(1) = 0$, and we let $t_1 = s_1$. And, having picked $t_1, \ldots, t_n$, we pick t_{n+1} as follows: if $m(n + 1) = m(n)$, we take as t_{n+1} the first s_k that has not been selected previously; and if $m(n + 1) > m(n)$, we take as t_{n+1} the first r_k that has not been previously selected. Now, by induction, it is readily seen that $m(n)$ is equal to the number of the r_k that have been selected in the first n steps. Since $m(1) = 0$ and the first element picked is s_1, the statement is true for $n = 1$. Assuming that the statement is true for n, i.e., $m(n)$ of the r_k have been selected in the first n steps, since we pick an s_k if $m(n+1) = m(n)$ and we pick an r_k if $m(n+1) > m(n)$, adding then 1 to the count, the statement is clearly true for $n + 1$.

Now, since

$$\frac{1}{n}\sum_{k=1}^{n}t_k = \frac{m(n)}{n}\frac{1}{m(n)}\sum_{k=1}^{m(n)}r_k + \frac{n-m(n)}{n}\frac{1}{n-m(n)}\sum_{k=1}^{n-m(n)}s_k,$$

it readily follows that

$$\lim_{n}\frac{1}{n}\sum_{k=1}^{n}t_k = \lambda r + (1-\lambda)s = t,$$

thus completing the proof of the first statement.

Finally, we claim that $\{t_k\}$ is u.d. if $\{r_k\}, \{s_k\}$ are u.d. To see this, let $[a,b] \subseteq [0,1]$. Given $\varepsilon > 0$, for an integer n, denote by m_n the number of the r_k among the first n terms of the sequence $\{t_k\}$, and pick N large enough so that for all $n \geq N$,

$$\left|\frac{1}{m_n}\sum_{k=1}^{m_n}\chi_{[a,b]}(r_k) - (b-a)\right| + \left|\frac{1}{n-m_n}\sum_{k=1}^{n-m_n}\chi_{[a,b]}(s_k) - (b-a)\right| \leq \varepsilon.$$

Then, since

$$\frac{1}{n}\sum_{k=1}^{n}\chi_{[a,b]}(t_k) - (b-a) = \frac{m_n}{n}\left(\frac{1}{m_n}\sum_{k=1}^{m_n}\chi_{[a,b]}(r_k) - (b-a)\right)$$

$$+ \frac{n-m_n}{n}\left(\frac{1}{n-m_n}\sum_{k=1}^{n-m_n}\chi_{[a,b]}(s_k) - (b-a)\right),$$

for $n \geq N$, it follows that

$$\left|\frac{1}{n}\sum_{k=1}^{n}\chi_{[a,b]}(t_k) - (b-a)\right| \leq \frac{m_n}{n}\varepsilon + \frac{n-m_n}{n}\varepsilon = \varepsilon,$$

and, therefore, $\{t_k\}$ is u.d. mod 1.   □

Theorem 19 *Let φ be a bounded function defined on I that is not Riemann integrable there. Then, given $\lambda \in \left(L_{\Pi_e}(\varphi), U_{\Pi_e}(\varphi)\right)$, there exists a u.d. sequence $\{z_k\}$ such that*

$$\lim_{n}\frac{1}{n}\sum_{k=1}^{n}\varphi(z_k) = \lambda.$$

Proof By Proposition 17, there are u.d. sequences $\{x_k\}$ and $\{y_k\}$ such that

$$L_{\Pi_e}(\varphi) = \lim_n \frac{1}{n} \sum_{k=1}^{n} \varphi(x_k), \quad \text{and,} \quad U_{\Pi_e}(\varphi) = \lim_n \frac{1}{n} \sum_{k=1}^{n} \varphi(y_k).$$

Consider now the sequences $\{r_k\} = \{\varphi(x_k)\}$ and $\{s_k\} = \{\varphi(y_k)\}$. Then, with $t = \lambda$ in Theorem 18, there is a dovetailing $\{t_k\}$ of $\{r_k\}$ and $\{s_k\}$ such that $\lim_n (1/n) \sum_{k=1}^{n} t_k = t$. Define now the sequence $\{z_k\}$ by the relation $\varphi(z_k) = t_k$, all k, and observe that

$$\lim_n \frac{1}{n} \sum_{k=1}^{n} \varphi(z_k) = \lim_n \frac{1}{n} \sum_{k=1}^{n} t_k = t = \lambda.$$

Moreover, we claim that $\{z_k\}$ is u.d. To see this, observe that each z_k is an x_h or a y_h, and that the number m_n of the x_h selected after n steps is the same as the number of the r_h in Theorem 18, and that the number $n - m_n$ of the y_h selected after n steps in Theorem 18 is the same as the number of the s_h, and so they are both infinite. We now repeat the proof that $\{t_k\}$ is u.d. in Theorem 18 with r_k replaced by x_k, s_k replaced by y_k, and t_k by z_k, and the result follows.   □

We also have this interesting result:

Theorem 20 *Let f, φ be real functions defined on $I = [0, 1]$, with f Riemann integrable and φ bounded. Then, if φ is Riemann integrable and $\{x_k\}$ is a u.d. mod 1 sequence, with $\mathcal{P}_n = \{I_k^n\}$ the partitions of I in Π_e, for arbitrary tags $c_k^n \in I_k^n$, all k, n, we have*

$$\lim_n \frac{1}{n} \sum_{k=1}^{n} \varphi(x_k) f(c_k^n) = \left(\int_I \varphi \right) \left(\int_I f \right). \tag{70}$$

On the other hand, if φ is not Riemann integrable on I, given λ in the interval $[L_{\Pi_e}(\varphi), U_{\Pi_e}(\varphi)]$, there exists a u.d. mod 1 sequence $\{z_k\}$ such that for $\mathcal{P}_n = \{I_k^n\}$ the partitions of I in Π_e, for arbitrary tags $c_k^n \in I_k^n$, all k, n, we have

$$\lim_n \frac{1}{n} \sum_{k=1}^{n} \varphi(z_k) f(c_k^n) = \lambda \int_I f.$$

Proof Assume first that $0 \le \varphi \le 1$ throughout I, put $\alpha_k = \varphi(x_k)$, all k, and let Φ be defined as in Example 3 for the sequence $\{\alpha_k\}$. Then by (69), $\alpha = \int_I \varphi$, and by (67),

$$\lim_n \sum_{k=1}^{n} f(c_k^{n,1})|I_k^{n,1}| = \lim_n \frac{1}{n} \sum_{k=1}^{n} \varphi(x_k) f(c_k^{n,1}) = \left(\int_I \varphi \right) \left(\int_I f \right). \tag{71}$$

Now, if $C = \{c_1^n, \ldots, c_n^n\}$ are arbitrary tags in I_k^n, by an argument similar to that in the Bliss' theorem,

$$\left| \frac{1}{n} \sum_{k=1}^{n} \varphi(x_k) f(c_k^n) - \frac{1}{n} \sum_{k=1}^{n} \varphi(x_k) f(c_k^{n,1}) \right| \leq \frac{1}{n} \sum_{k=1}^{n} \text{osc } (f, I_k^n),$$

which by (6) in Proposition 1 tends to 0 as $n \to \infty$, and, consequently, by (71), (70) holds in this case.

Moreover, if $\varphi \geq 0$, since Riemann integrable functions are bounded, pick M such that $\varphi_1 = \varphi/M \leq 1$. Then, (71) applied to φ_1 gives the conclusion, since M cancels out. Thus the conclusion holds for φ nonnegative.

Finally, if φ changes signs, consider φ^+ and φ^-. Then (70) holds for φ^+ and φ^-, and since $\varphi = \varphi^+ - \varphi^-$, and $\int_I \varphi = \int_I \varphi^+ - \int_I \varphi^-$, it also holds for an arbitrary φ, thus completing the proof in this case.

The proof of the second statement follows along similar lines once we invoke Theorem 19 and identify the sequence $\{z_k\}$. $\square$

Chapter 5
The Pattern and Uniform Integrals

In this chapter, we discuss the pattern and uniform integrals. Whereas the original applications of the pattern integrals were limited to summability methods [45], we are interested in sequences more general than patterns, the modified Π-Riemann sums being the natural setting for these results.

A double sequence $\{\alpha_{k,n}\}$ defined for $1 \leq k \leq n$, $n = 1, 2, \ldots$, is said to be a *variable pattern* if $\alpha_{k,n} = 0$ or $\alpha_{k,n} = 1$ for all k, n. As a particular instance of Example 2, the mapping Φ defined there on the subintervals I_k^n of the family $\Pi_e = \{\mathcal{P}_n\}$ assumes the value $\Phi(I_k^n) = I_k^n$ if $\alpha_{k,n} = 1$ and $\Phi(I_k^n) = \emptyset$ if $\alpha_{k,n} = 0$, $1 \leq k \leq n$, $n = 1, 2, \ldots$ Then the summation in (1) becomes

$$S(f, \mathcal{P}_n, C) = \sum_{k, \alpha_{k,n}=1} f(c_k^n) \, |I_k^n|$$

and extends over the variable pattern determined by the $\alpha_{k,n}$ for each level n. By Proposition 16, provided (56) holds, the Π_e-Riemann integral of f associated with this variable pattern is

$$u_{\Pi_e}(f) = \alpha \int_I f.$$

This result contains Theorem 3.1 in [17].

As for the *patterns*, i.e., prescribed subsets P of $\mathbb{N}$, [17], the sums associated with P assume the form

$$(P) \sum_k f(c_k^n) \, |I_k^n| = \sum_{1 \leq k \leq n, k \in P} f(c_k^n) \, |I_k^n|, \tag{72}$$

A. Torchinsky, *A Modern View of the Riemann Integral*,
Lecture Notes in Mathematics 2309, https://doi.org/10.1007/978-3-031-11799-2_5

where the summation is restricted to the subset P of $\mathbb{N}$. We say that $(P) \int_I f$ is the *pattern integral* of f on I if the limit

$$\lim_n (P) \sum_k f(c_k^n) |I_k^n| = (P) \int_I f$$

exists [17].

Following [17], we say that P is a *fixed pattern* if it can be characterized uniquely by a dyadic number

$$t = 0.\alpha_1 \alpha_2 \ldots \alpha_k \ldots$$

where if $\alpha_k = 1$, the kth term of the Riemann sum is to be taken, and if $\alpha_k = 0$, the kth term of the Riemann sum is to be omitted.

Let $\{\alpha_n\}$ be a sequence of nonnegative real numbers such that $\alpha_n = 0$, or $\alpha_n = 1$ for all n. Referring to Example 3, the mapping Φ defined on the subintervals I_k^n of the family $\Pi_e = \{\mathcal{P}_n\}$ assumes the values $\Phi(I_k^n) = I_k^{n,1} = I_k^n$, if $\alpha_k = 1$, and $\Phi(I_k^n) = \emptyset$ if $\alpha_k = 0$. The principal theorem in [17] is then a particular case of (67) above. To the point, we have:

Principal Theorem *Let $I = [a, b]$ be a finite interval in $\mathbb{R}$, and suppose that the pattern P is characterized by a given t such that*

$$\lim_n \frac{1}{n} \sum_{k=1}^n \alpha_k = \alpha.$$

Let f be Π_e-Riemann integrable on I. Then, the (P)-Riemann integral of f exists, and we have

$$(P) \int_I f = \alpha \int_I f. \tag{73}$$

Note that the pattern integral of a function f may exist (in the sense that the sums in (72) have a limit), even if the limit in (66) does not exist, and therefore, (73) cannot hold; an instance of this is given in Example 1 in [45]. On the other hand, by the Law of Large Numbers, the average sums of the terms of a sequence consisting of 0's and 1's converge to $1/2$ for almost all such sequences, i.e., the Cesàro means of order 1 of such sequences are $1/2$ for almost all those sequences. From this, we conclude that for fixed patterns, almost all pattern integrals of Riemann integrable functions are equal to $1/2$ the corresponding Riemann integrals.

The *Riemann sum operators* $M_n f(x)$, where f is a real, periodic function of period 1, are a particular instance of the pattern integral. They are defined on I as

$$M_n f(x) = \frac{1}{n} \sum_{k=1}^n f\left(x + \frac{k}{n}\right), \quad x \in \mathbb{R}.$$

When $x = 0$, $M_n f(0)$ is the pattern integral of f where all the α_k are equal to 1, and, therefore, by the principal theorem, we have

$$\lim_n M_n f(0) = (P) \int_I f = \int_I f.$$

By the periodicity of f, the same is true for all $x \in \mathbb{R}$, and therefore,

$$\lim_n M_n(f)(x) = \int_I f, \quad \text{all } x \in \mathbb{R}.$$

This convergence statement, which is not true for Lebesgue integrable functions, has interesting number-theoretic aspects [83].

We consider next an integral that is defined in terms of the usual limits. We say that f is *uniform integrable* on $I = [a, b]$ with *uniform integral* equal to $(U) \int_I f$ provided that the limit

$$\lim_{c \to 0^+} c \sum_{k=[a/c]+1}^{[b/c]} f(ck) = (U) \int_I f$$

exists [85, 106]. As we shall see below, the uniform integral extends the Riemann integral properly: any odd function is (U) integrable in a symmetric interval about the origin, with integral $= 0$. A less dramatic example is the function f defined on $[0, 1]$ by $f(x) = x$ if x is rational and $f(x) = 1/2$ otherwise, which is uniform integrable, yet fails to be Riemann integrable on $[0, 1]$.

Although subinterval integrability may fail for a function that is uniform, but not Riemann, integrable [106], the following is true. Let $d \in (a, b)$, and $c > 0$. Since

$$c \sum_{k=[a/c]+1}^{[d/c]} f(ck) + c \sum_{k=[d/c]+1}^{[b/c]} f(ck) = c \sum_{k=[a/c]+1}^{[b/c]} f(ck),$$

if the limit of any two of the above sums exists as $c \to 0^+$, the limit of the third sum also exists. Thus, in particular, if f is uniform integrable on $[a, b]$ and on $[d, b]$, then f is uniform integrable on $[a, d]$ and

$$(U) \int_{[a,d]} f + (U) \int_{[d,b]} f = (U) \int_{[a,b]} f. \tag{74}$$

And, as stated above, the class of uniform integrals contains the class of Riemann integrable functions. Indeed, we have:

Theorem 21 *Suppose that f is Riemann integrable on $I = [a, b]$. Then f is uniform integrable on I, and $\int_I f = (U) \int_I f$.*

Proof Given a small $c > 0$, let $m = [a/c]$ and $n = [b/c]$, and write

$$c \sum_{k=[a/c]+1}^{[b/c]} f(ck) = c \sum_{k=m+1}^{n} f(ck) = c \sum_{k=1}^{n-m} f(c(m+k)).$$

With $J_k = [a + (k-1)c, a + kc]$, $1 \le k \le (n-m)$, observe that

$$c(m+k) \in J_k = [a + (k-1)c, a + kc], \quad k = 1, \ldots, n-m,$$

and, in particular, for $k = n - (m+1)$, we have

$$J_k = [a + (n-m-2)c, a + (n-m-1)c],$$

where, since $nc \le b$ and $a < (m+1)c$, we have $a + (n-m-1)c \le a+b-a = b$. Also, if $J = [a + (n - (m+1))c, b]$, since $mc \le a$ and $b < (n+1)c$, we have $|J| \le 2c$. Write now the integral as

$$\int_I f = \sum_{k=1}^{n-(m+1)} \int_{I_k} f + \int_J f,$$

and the sum as

$$c \sum_{k=1}^{n-m} f(c(m+k)) = c \sum_{k=1}^{n-(m+1)} f(c(m+k)) + |J| f(b) + \big(cf(nc) - |J| f(b)\big)$$

$$= \sigma + \big(cf(nc) - |J| f(b)\big),$$

where, with M_f a bound for f, since $nc \sim b$, we have $\big|cf(nc) - |J| f(b)\big| \le M_f(c + 2c)$, which tends to 0 as $c \to 0^+$.

Now observe that $\big|\int_I f - \sigma\big|$ is dominated by

$$\left| \sum_{k=1}^{n-(m+1)} \int_{I_k} f - c \sum_{k=1}^{n-(m+1)} f(c(m+k)) \right| + \left| \int_J f - |J| f(b) \right|$$

$$\le \sum_{k=1}^{n-(m+1)} \int_{I_k} \big| f - f(c(m+k)) \big| + \int_J \big| f - f(b) \big|$$

$$\le \sum_{k=1}^{n-(m+1)} \mathrm{osc}\,(f, I_k)\,|I_k| + \mathrm{osc}\,(f, J)\,|J|. \tag{75}$$

Since for each $c > 0$, $\mathcal{P} = \{J_1, \ldots, J_{n-m+1}, J\}$ is a partition of $[a, b]$ with $\|\mathcal{P}\| \to 0$ as $c \to 0^+$, by (6) in Proposition 1, the expression on the right-hand side of (75) tends to 0 as $c \to 0^+$. Therefore,

$$\lim_{c \to 0^+} c \sum_{k=[a/c]+1}^{[b/c]} f(ck) = \int_I f,$$

and we have finished.   □

The connection of the uniform integral with the geometry of numbers is examined in [106]. We will examine the rich connection with improper integrals next.

Chapter 6
The Improper and Dominated Integrals

In this chapter, we will discuss the improper and dominated integrals. We have considered so far the Riemann integral of bounded functions defined on finite closed intervals, but there are instances when the function we are interested in integrating does not satisfy one or both of these conditions. The two basic types of such functions are those that are unbounded in the neighbourhood of a point in the interval, say the left endpoint, where it tends to ∞, and those whose domain of integration is infinite and are Riemann integrable on any finite interval included in that domain. We will incorporate these functions into our framework by defining their integrals as limits of Riemann integrals. This will then involve two limits, a limit of Riemann sums to define the Riemann integrals, followed by a limit of Riemann integrals. These integrals will be called *improper Riemann integrals*. Functions that satisfy the former condition will be covered by improper integrals of the first type and those that satisfy the later condition by improper integrals of the second type.

Concerning improper integrals, a couple of general observations about the results we discuss in detail below. The notion of dominated integration addresses the question as to when a function has a positive steadily decreasing majorant with a convergent improper integral of the first type. And, a most interesting case for integrals of the second type occurs when the set of partition intervals of $[0, \infty)$ is that divided by the points $0 < h < 2h < \ldots$, where h is a parameter that tends to 0, i.e., $x_n = nh$, [12].

We also explore the relation between uniform and improper integrals and note the simplicity of computation when dealing with uniform integrals, and this is true in other contexts as well. Quadrature rules converge to the integral of the function when the function is Riemann integrable, but in the case of the improper Riemann integral this connection is obscured by the double limiting process involved. The dominated integral, which is defined by a single limit, allows for this connection.

© The Author(s), under exclusive license to Springer Nature Switzerland AG 2022
A. Torchinsky, *A Modern View of the Riemann Integral*,
Lecture Notes in Mathematics 2309, https://doi.org/10.1007/978-3-031-11799-2_6

6.1   Improper Integrals of the First Type

For functions f of the first type defined on an interval $I = (a, b]$, we say that the improper Riemann integral of f converges and has value $\int_{[a^+,b]} f$ provided that f is Riemann integrable on $[a + \varepsilon, b]$ for all $\varepsilon > 0$, with limit

$$\lim_{\varepsilon \to 0^+} \int_{[a+\varepsilon,b]} f = \int_{[a^+,b]} f. \tag{76}$$

It is clear that if the integral of f exists in the ordinary sense, the improper integral of f converges and both have the same value. And, that the improper Riemann integral does not share some of the limitations of the Lebesgue integral. Consider, e.g., $f(x) = x^2 \sin(\pi/x^2)$, $x \neq 0$, and $f(0) = 0$. Then,

$$\int_{[\varepsilon,1]} f' = -\varepsilon^2 \sin(\pi/\varepsilon^2) \to 0 \quad \text{as } \varepsilon \to 0^+,$$

and so the improper Riemann integral of f' converges, whereas, since the Lebesgue integral of $|f'|$ on $[0, 1]$ is ∞, f' is not Lebesgue integrable.

In what follows, we let $I = [0, 1]$. We are interested in defining an improper integral by means of a single limit. It is clear that by picking tags $\{c_1^n\}$ where c_1^n is an arbitrary point in the interval $(0, 1/n)$ in the partitions of Π_e, the usual definition of the Riemann integral fails. Bromwich and Hardy considered continuous functions and suggested two choices for the tags, one for a function of each type [12]. For functions of the first type and partitions $\mathcal{P}_n$ of I in Π_e, their choice is $c_k^n = k/n$, for $k = 1, \ldots, n$, and the corresponding approximating expression is the right Riemann sum of f,

$$B_n = S_R(f, \mathcal{P}_n) = \frac{1}{n} \sum_{k=1}^{n} f(k/n), \quad \text{all } n. \tag{77}$$

Generally speaking, the Riemann sums will be restricted to those sums where the tag in the first interval is chosen not too near 0, and the first interval of the partition is not too small compared to the mesh of the partition, which measures the largest length among the subintervals of the partition.

We will first consider the case when f is nonnegative. Then it is readily seen that for the limit in (76) to exist, it suffices that it exists for a fixed sequence going to 0. Moreover, for positive functions that tend steadily to ∞ as $x \to 0^+$, we have:

Theorem 22 *Let f be a positive, decreasing function defined on $(0, 1]$ that increases to ∞ as x decreases to 0. Then the following are equivalent:*

(i) $\lim_{x \to 0^+} x f(x) = 0$, *and f is (U) integrable on I.*
(ii) *The improper integral $\int_{[0^+,1]} f$ converges.*

(iii) $\lim_{x \to 0+} x f(x) = 0$, and $\lim_n S_n = S$ exists when $\lim_n \|\mathcal{P}_n\| = 0$, where the Riemann sums $S_n = S(f, \mathcal{P}_n, C_n)$ correspond to partitions $\mathcal{P}_n$ and tags $C_n = \{c_1^n, \ldots, c_{m_n}^n\}$ such that for some constants $0 < \lambda, \mu < 1$, $c_1^n > \mu x_1^n$, and $x_1^n \geq \lambda \|\mathcal{P}_n\|$ for all n.

(iv) $\lim_{x \to 0+} x f(x) = 0$, and $\lim_n A_n = A$ exists, where

$$A_n = \frac{1}{n} \sum_{k=1}^{n} f(c_k^n),$$

and the tags $C = \{c_1^n, \ldots, c_n^n\}$ are such that, with $M \geq 1$ fixed, $1/Mn \leq c_1^n \leq 1/n$, and $(k-1)/n \leq c_k^n \leq k/n$, for $k = 2, \ldots, n$.

Furthermore, we have $(U) \int_{[0,1]} f = S = A = \int_{[0^+,1]} f$.

Proof (i) implies (ii). For (small) $c > 0$, put $[1/c] = n$. Note that by the monotonicity of f,

$$-\int_{[ck,c(k+1)]} f \leq -cf(c(k+1)), \qquad 1 \leq k \leq (n-1),$$

and write

$$c \sum_{k=1}^{n} f(ck) - \int_{[c,1]} f$$

$$= \left(c \sum_{k=1}^{n-1} f(ck) - \sum_{k=1}^{n-1} \int_{[ck,c(k+1)]} f \right) + \left(cf(cn) - \int_{[cn,1]} f \right)$$

$$= C_n + D_n, \tag{78}$$

say.

We will consider D_n first. Since $nc \sim 1$ and f is bounded near 1, $cf(cn) \to 0$ as $c \to 0^+$. Also, since $|[cn, 1]| \leq c$ and f is bounded on that interval, $|\int_{[cn,1]} f| \to 0$ as $c \to 0^+$. Hence, $\lim_n |D_n| = 0$.

As for C_n, combining the sums, it follows that

$$C_n = \sum_{k=1}^{n-1} \int_{[ck,c(k+1)]} (f(ck) - f), \tag{79}$$

where, by the monotonicity of f, each summand in (79) is nonnegative. Moreover, by the monotonicity of f and by telescoping it, it follows that

$$C_n \leq \sum_{k=1}^{n-1} \Big(cf(ck)) - cf(c(k+1)) \Big) = c(f(c) - f(nc)),$$

which tends to 0 since both $cf(c)$ and $cf(nc)$ tend to 0 as $c \to 0^+$.
    Hence,

$$\limsup_{c \to 0^+} \Big| c \sum_{k=1}^{n} f(ck) - \int_{[c,1]} f \Big| \leq \limsup_{c \to 0^+} (C_n + |D_n|) = 0, \tag{80}$$

and the left-hand side of (80) tends to 0 as $c \to 0^+$.
    Thus, since

$$\Big| \int_{[c,1]} f - (U) \int_I f \Big| \leq \Big| \int_{[c,1]} f - c \sum_{k=1}^{n} f(ck) \Big| + \Big| c \sum_{k=1}^{n} f(ck) - (U) \int_I f \Big|,$$

where the first summand goes to 0 as $c \to 0^+$ by (80) and the second summand goes to 0 as $c \to 0^+$ since the uniform integral of f converges, it follows that the improper Riemann integral of f converges to $(U) \int_I f$.
    (ii) implies (iii). First note that if the improper integral of f converges, for $0 < x < 1$ it follows that

$$xf(x) \leq 2 \int_{[x/2,x]} f = 2 \Big(\int_{[x,1]} f - \int_{[x/2,1]} f \Big) \to 0 \quad \text{as } x \to 0^+.$$

Let now $\mathcal{P}_n = \{I_1^n, \ldots, I_{m_n}^n\}$ be partitions of $[0, 1]$ with corresponding tags C_n, and, specifically, let $C_{n,1} = \{x_1^n, \ldots, x_{m_n}^n\}$ and $C_{n,2} = \{c_1^n, x_1^n, \ldots, x_{m_n-1}^n\}$. Since f is nonincreasing, $f(x_1^n) \leq f(c_1^n)$, and $f(x_k^n) \leq f(c_k^n) \leq f(x_{k-1}^n)$ for $2 \leq k \leq m_n$, for all n, and, consequently,

$$S(f, \mathcal{P}_n, C_{n,1}) \leq S(f, \mathcal{P}_n, C_n) \leq S(f, \mathcal{P}_n, C_{n,2}). \tag{81}$$

We claim that $S(f, \mathcal{P}_n, C_{n,2}) - S(f, \mathcal{P}_n, C_{n,1}) \to 0$ as $\|P_n\| \to 0$. To see this, note that $S(f, \mathcal{P}_n, C_{n,2}) - S(f, \mathcal{P}_n, C_{n,1})$ is equal to

$$\Big(f(c_1^n) x_1^n - f(x_1^n) x_1^n \Big) + \sum_{k=2}^{m_n} \Big(f(x_{k-1}^n) - f(x_k^n) \Big) \Big(x_k^n - x_{k-1}^n \Big),$$

where the first summand is bounded by $f(c_1^n) x_1^n \le f(c_1^n) c_1^n/\mu$, and the second summand is bounded by

$$\|\mathcal{P}_n\| \sum_{k=2}^{m_n} \left(f(x_{k-1}^n) - f(x_k^n) \right) \le \|\mathcal{P}_n\| f(x_1^n) \le f(x_1^n) x_1^n/\lambda.$$

Hence, given $\varepsilon > 0$, since $\lim_{x \to 0^+} x f(x) = 0$, we can choose $\eta_0 > 0$ such that with $0 < c_1^n \le x_1^n \le \|\mathcal{P}_n\|$,

$$S(f, \mathcal{P}_n, C_{n,2}) - S(f, \mathcal{P}_n, C_{n,1}) \le f(c_1^n) c_1^n/\mu + f(x_1^n) x_1^n/\lambda \le \varepsilon/2, \qquad (82)$$

for all partitions with $\|\mathcal{P}_n\| \le \eta_0$.

Now, η_0 can be chosen so that, if $0 < x_1^n < \eta_0$, with $I = \int_{[0^+,1]} f$, we have

$$I - \varepsilon/2 < \int_{[x_1^n,1]} f \le I. \qquad (83)$$

Since f is decreasing, we also have

$$S(f, P_n, C_{n,1}) \le \int_{[x_1^n,1]} f \le S(f, \mathcal{P}_n, C_{n,2}), \qquad (84)$$

and, consequently, by (82) and (84),

$$0 \le \int_{[x_1^n,1]} f - S(f, P_n, C_{n,1}) \le \varepsilon/2. \qquad (85)$$

Hence, by (83) and (85),

$$\left| I - S(f, P_n, C_{n,1}) \right| \le \left| I - \int_{[x_1^n,1]} f \right| + \left| \int_{[x_1^n,1]} f - S(f, P_n, C_{n,1}) \right| \le \varepsilon.$$

This implies that $S(f, P_n, C_{n,1}) \to I$ as $\|P_n\| \to 0$. Now, since by (82), $S(f, P_n, C_{n,2}) - S(f, P_n, C_{n,1}) \to 0$, it follows that also $S(f, P_n, C_{n,2}) \to I$ as $\|P\| \to 0$, and by (81), the proof is finished.

(iii) implies (iv). Take $\mu = 1/M$, $\lambda = 1$. The conclusion also applies to the B_ns.

(iv) implies (i). Let $\lceil 1/c \rceil = n$. Then, $n = \lceil 1/c \rceil \le 1/c < (n+1)$, which implies that for $1 \le k \le n$, we have $k/(n+1) < ck \le k/n$. Now, since it readily follows that in this range of k, n, we have $(k-1)/n \le k/(n+1)$, $ck \in I_k^n = [(k-1)/n, k/n]$, all $k \le n$. Thus, if we let

$$A_n = \frac{1}{n} f(1/n) + \frac{1}{n} \sum_{k=2}^{n} f(ck),$$

the A_n satisfy the conditions in the definition in (iv).

Now,

$$|A_n - c \sum_{k=1}^{n} f(ck)| \le \left|\frac{1}{n}f(1/n) - cf(c)\right| + \left|\left(\frac{1}{n} - c\right)\right| \left|\sum_{k=2}^{n} f(ck)\right|. \tag{86}$$

Moreover, since $1 - cn \le c$, it readily follows that

$$0 \le \frac{1}{n} - c = \frac{1}{n}\left(1 - nc\right) \le \frac{c}{n} \le \frac{1}{n^2},$$

and the second summand above is bounded by

$$\frac{1}{n}|A_n| + \frac{1}{n^2}|f(1/n)|,$$

which, by assumption, and since A_n being convergent is bounded, goes to 0 as $n \to \infty$. Also the first summand in (86) goes to 0 by assumption.

Finally,

$$\left|A - c \sum_{k=1}^{n} f(ck)\right| \le \left|A - A_n\right| + \left|A_n - c \sum_{k=1}^{n} f(ck)\right|,$$

where the first term goes to 0 by the convergence of the A_n, and the second goes to 0 by the computation. Thus, f is uniform integrable on I with $(U)\int_I f = A$. □

Theorem 22 is not necessarily true when the monotonicity assumption is relaxed. Indeed, suppose that $f(x) = 2^n$, for $x = 2^{-n}$, all $n \ge 0$. Then the sum $2^{-n}\sum_{k=1}^{2^n} f(k/2^n)$ includes the term $(1/2^n)f(1/2^n) = 1$, and so it is ≥ 1. Hence, $(1/n)\sum_{k=1}^{n} f(k/n)$ cannot converge to a limit less than 1 as $n \to \infty$. But, it is easy to define a function f (graphically or otherwise) that is positive and continuous for $x > 0$, assumes the values assigned above, and yet $\int_{[0^+,1]} f < \eta$, for an arbitrary η.

It is possible however to remove the restrictions of Theorem 22 by assuming that f is majorized throughout $(0, 1]$ by a monotone function that has a convergent improper integral [28]. Specifically:

Theorem 23 *Let f be a function defined on $(0, 1]$ such that $|f(x)| \le F(x)$ throughout $(0, 1]$, where F is a monotone decreasing function whose improper integral converges. Then f is absolutely improper Riemann integrable on $[0, 1]$, and if Π is a class of partitions of $[0, 1]$ containing partitions $\mathcal{P}$ with arbitrarily small mesh $\|\mathcal{P}\|$ such that*

$$\lim_{\|\mathcal{P}\| \to 0} S_\Pi(F, \mathcal{P}, C) = \int_{[0^+,1]} F,$$

then for the same family Π *of partitions* $\mathcal{P}$ *of* $[0, 1]$,

$$\lim_{\|\mathcal{P}\|\to 0} S_{\Pi}(f, \mathcal{P}, C) = \int_{[0^+,1]} f.$$

Proof Given $\varepsilon > 0$, let $\delta > 0$ be such that

$$\int_{[0^+,\delta]} F \leq \varepsilon/6. \tag{87}$$

Then, for $0 < a < b < \delta$, we have

$$\left| \int_{[a,b]} f \right| \leq \int_{[a,b]} |f| \leq \int_{[0^+,\delta]} F \leq \varepsilon/6,$$

and, consequently, f is improper, and absolutely improper, Riemann integrable on $[0, 1]$.

Write now

$$I = \int_{[0^+,1]} f = \int_{[0^+,\delta]} f + \int_{[\delta,1]} f = I' + I'',$$

and

$$J = \int_{[0^+,1]} F = \int_{[0^+,\delta]} F + \int_{[\delta,1]} F = J' + J'',$$

say.

Let now $\mathcal{P} = \{I_k\}$, $1 \leq k \leq n$, be a partition of $[0, 1]$ in Π with $\|P\| \leq \eta_0$. There are two possibilities, δ is an endpoint of an interval in $\mathcal{P}$ or δ is an interior point of one of the intervals of $\mathcal{P}$. We will only do the latter case, since the proof of the former case follows along similar, simpler, lines. So, if δ is an interior point of $I_{k_0} = [x_{k_0,l}, x_{k_0,r}] \in \mathcal{P}$, say, put $I_{k_0,l} = [x_{k_0,l}, \delta]$, $I_{k_0,r} = [\delta, x_{k_0,r}]$, and let $Q = \{I_1, \ldots, I_{k_0-1}, I_{k_0,l}, I_{k_0,r}, I_{k_0+1}, \ldots, I_n\}$ be a refinement of $\mathcal{P}$, with the tag set C_Q consisting of the tag set of $\mathcal{P}$ with an additional point; Q is not necessarily in Π. Now, since $|S_{\Pi}(F, \mathcal{P}, C) - S(F, Q, C_Q)| \to 0$ with $\|\mathcal{P}\| \to 0$, we can choose $\eta_1 \leq \eta_0$ such that, whenever $\|P\| \leq \eta_1$,

$$|S(F, Q, C_Q) - J| \leq \varepsilon/6. \tag{88}$$

Note that $\mathcal{P}' = \{I_1, \ldots, I_{k_0,l}\}$ is a partition of the interval $[0, \delta]$ with tag set $C' = \{c_1^n, \ldots, c_{k_0-1}^n, c_{k_0,l}^n\}$, and, similarly, $P'' = \{I_{k_0,r}, \ldots, I_n\}$ is a partition of the interval $[\delta, 1]$ with tag set $C'' = \{c_{k_0,r}^n, c_{k_0+1}^n, \ldots, c_{m_n}^n\}$.

Now, since I'' and J'' are ordinary Riemann integrals, we can pick η_0 such that, if $\|P''\| \leq \eta_0$,

$$|S(f,\mathcal{P}'',C'') - I''| < \varepsilon/6, \quad |S(F,\mathcal{P}'',C'') - J''| < \varepsilon/6. \tag{89}$$

Since

$$S(F,\mathcal{P}',C') = \big(S(F,Q,C_Q) - J\big) - \big(S(F,\mathcal{P}'',C'') - J''\big) + J'$$

by (88), (87), and (89), it follows that

$$S(F,\mathcal{P}',C') \leq \big|S(F,Q,C_Q) - J\big| + \big|S(F,\mathcal{P}'',C'') - J''\big| + \big|J'\big| \leq \varepsilon/2,$$

which implies that $|S(f,\mathcal{P}',C')| \leq S(F,\mathcal{P}',C') \leq \varepsilon/2$.

Furthermore,

$$|S(f,Q,C_Q) - I| \leq |S(f,\mathcal{P}',C') - I'| + |S(f,\mathcal{P}'',C'') - I''|$$

$$\leq |S(f,\mathcal{P}',C')| + |I'| + |S(f,\mathcal{P}'',C'') - I''|,$$

where as we just have seen, the first summand above is estimated by $\varepsilon/2$, the second summand can be made not to exceed $\varepsilon/6$ on account of the proper Riemann integrability, and the last summand can be made $\leq \varepsilon/6$, by the usual Riemann integrability of the function. Thus, the expression is bounded by $5\,\varepsilon/6$.

Finally, since $|S(f,Q,C_Q) - S_\Pi(f,\mathcal{P},C)| \to 0$ with $\|\mathcal{P}\|$, it follows that $|S_\Pi(f,\mathcal{P},C) - I| \leq \varepsilon$, for $\|P\|$ sufficiently small, and the proof is complete. $\square$

On account of Theorem 23, it is of interest to determine under what conditions functions f defined on $(0, 1]$ have a nonnegative decreasing improper integrable function as a majorant, for then the results of Theorem 22 will transfer to f; we will return to this in Theorem 25. In the meantime, not to lose the thread, we will round up the other results.

Our next result invokes the notion of *oscillation along a partition*, defined as follows. Given $c > 0$, and $n = [1/c]$, let

$$V(f,c) = c \sum_{k=1}^{n-1} \text{osc}\,(f, [ck, c(k+1)]).$$

We then have:

Theorem 24 *Let f be defined and finite on $(0, 1]$, and assume that f satisfies*

$$\lim_{c \to 0^+} V(f,c) = 0. \tag{90}$$

Then, the improper Riemann integral of f converges iff the uniform integral of f exists, and in that case $\int_{[0^+,1]} f = (U)\int_I f$.

Proof First observe that $\lim_{c \to 0^+} cf(c) = 0$. Indeed, since

$$f(c) = \sum_{k=1}^{n-1} (f(ck) - f(c(k+1))) + f(nc),$$

it readily follows that $c|f(c)| \leq V(f,c) + c|f(nc)|$, which, by (90) and since $nc \sim 1$, tends to 0 as $c \to 0^+$.

We do the necessity first. Note that, with $[1/c] = n$,

$$\left| c \sum_{k=1}^{n} f(ck) - \int_{[0^+,1]} f \right|$$

$$\leq \left| c \sum_{k=1}^{n} f(ck) - \int_{[c,1]} f \right| + \left| \int_{[c,1]} f - \int_{[0^+,1]} f \right|, \tag{91}$$

where on account of the existence of the improper integral of f the second terms go to 0 as $c \to 0^+$.

As for the sum in the first term, as in (78), it can be written as $C_n + D_n$. Note that for $x \in [ck, c(k+1)]$, $|f(ck) - f(x)| \leq \operatorname{osc}(f, [ck, c(k+1)])$, and, consequently, by (79),

$$|C_n| \leq c \sum_{k=1}^{n-1} \operatorname{osc}(f, [ck, c(k+1)]) = V(f,c),$$

which by assumption goes to 0 as $c \to 0^+$. And, $D_n \to 0$ as above.

Combining these estimates, it follows that

$$\lim_{c \to 0^+} c \sum_{k=1}^{n} f(ck) = \int_{[0^+,1]} f,$$

and, therefore, the uniform integral of f exists and is equal to the improper integral of f.

The proof of the sufficiency follows along similar lines. Consider, with $[1/c] = n$,

$$\left| \int_{[c,1]} f - (U) \int_{I} f \right|$$

$$\leq \left| \int_{[c,1]} f - c \sum_{k=1}^{n} f(ck) \right| + \left| c \sum_{k=1}^{n} f(ck) - (U) \int_{I} f \right|,$$

where the second terms go to 0 as $c \to 0^+$ on account of the existence of the uniform integral of f. As for the first term, it is precisely $C_n + D_n$ above, and so it tends to 0 as $c \to 0^+$. Therefore, it follows that $\lim_{c \to 0^+} \int_{[c,1]} f = (U) \int_I f$, and, therefore, the improper integral of f exists and is equal to the uniform integral of f.

$\square$

The necessity of Theorem 24 extends a result of Wintner, stated in terms of the total variation of f, [108]. As for the sufficiency with the assumption (90) removed, Wintner notes that it is quite deep as it contains results in Lambert summability as well as the prime number theorem [48, 88, 108]. The convergence of the improper integral required for the latter proof relies on a series using the Möebius function; a more elementary proof of this result remains a tantalizing challenge. A preliminary result in this direction is:

Proposition 18 *Let f be defined on $(0, 1]$, Riemann integrable on $[r, 1]$ for $0 < r < 1$, and uniform integrable on I, and let B_n be given by (77). Then, $\lim_n B_n = (U) \int_I f$. Moreover, if $\lim_{\varepsilon \to 0^+} (U) \int_{[0,\varepsilon]} f = 0$, the improper integral $\int_{[0^+,1]} f$ also converges to $(U) \int_I f$.*

Proof With ε a small positive number < 1, write

$$B_n = \frac{1}{n} \sum_{k=1}^{[\varepsilon n]} f(k/n) + \frac{1}{n} \sum_{k=[\varepsilon n]+1}^{n} f(k/n).$$

Now, since f has a uniform integral on $[\varepsilon, 1]$, being Riemann integrable there, and by assumption on I, by (74), f has a uniform integral on $[0, \varepsilon]$, and the limit

$$\lim_{c \to 0^+} c \sum_{k=1}^{[\varepsilon/c]} f(ck) = (U) \int_{[0,\varepsilon]} f \tag{92}$$

exists. Therefore, for every subsequence of $c \to 0^+$, the limit will also exist and be equal to the limit. Whence with the choice $c = 1/n, 1/(n+1), \ldots$, for n large, (92) gives

$$\lim_n \frac{1}{n} \sum_{k=1}^{[\varepsilon n]} f(k/n) = (U) \int_{[0,\varepsilon]} f. \tag{93}$$

The second summand leading to (74) needs a small adjustment. Let j_ε denote the largest integer j such that $j \leq n(1 - \varepsilon)$ and observe that

$$\frac{[\varepsilon n] + j}{n} \in \left[\varepsilon + \frac{(j-1)}{n}, \varepsilon + \frac{j}{n}\right] \subset I, \quad j \leq j_\varepsilon.$$

Let $I_\varepsilon \subset I$ and c_ε^n denote the interval and the tag $c_\varepsilon^n \in I_\varepsilon$,

$$I_\varepsilon = \left[\varepsilon + \frac{j_\varepsilon}{n}, 1 \right], \qquad c_\varepsilon^n = \varepsilon + \frac{j_\varepsilon}{n},$$

respectively. Then,

$$\frac{1}{n} \sum_{k=[n\varepsilon]+1}^{n} f(k/n) = \frac{1}{n} \sum_{k=[n\varepsilon]+1}^{n-1} f(k/n) + \frac{1}{n} f(1) \pm |I_N| \, f(c_N^n),$$

where

$$\sum_{k=[n\varepsilon]+1}^{n-1} f(k/n) + |I_N| \, f(c_N^n)$$

is a Riemann sum of f on $[\varepsilon, 1]$, and since $|I_N| \leq 1/n$,

$$\lim_n \frac{1}{n} |f(1)| + |I_N| \, |f(c_N^n)| = 0.$$

Hence,

$$\lim_n \frac{1}{n} \sum_{k=[n\varepsilon]+1}^{n} f(k/n) = \int_{[\varepsilon,1]} f. \tag{94}$$

Therefore, combining (93) and (94), it follows that

$$B = \lim_n B_n = (U) \int_{[0,\varepsilon]} f + \int_{[\varepsilon,1]} f = (U) \int_{[0,1]} f.$$

Furthermore,

$$\left| B - \int_{[\varepsilon,1]} f \right| \leq \left| (U) \int_{[0,\varepsilon]} f \right| \to 0,$$

and the proof is finished.   □

6.2   The Dominated Integral

We proceed now to the numerical evaluation of improper integrals of the first type by a single limit by means of the dominated integral. Let f be defined on $(0, 1]$. We say that f is *dominantly integrable* if there is a real number D with the following property: For every $\varepsilon > 0$, there exist $0 < \delta, \eta < 1$ such that

$$\left| \sum_{k=1}^{n} f(c_k)(x_k - x_{k-1}) - D \right| \le \varepsilon \tag{95}$$

whenever $0 < x_0 < x_1 < \ldots < x_n = 1$, $x_0 < \eta$, the tags $c_k \in I_k = [x_{k-1}, x_k]$, and $x_{k-1} > (1 - \delta) x_k$ for $k = 1, \ldots, n$.

It is clear that if such a D exists, it is unique; D is called the *dominated integral* of f. Also, the condition $x_{k-1} > (1 - \delta) x_k$ is equivalent to $|I_k| = x_k - x_{k-1} \le \delta x_k$, all k, and there is an equivalent definition involving an infinite sum in (95), [59].

An important property of these functions is that they have a convergent improper Riemann integral on $[0, 1]$. To see this, we begin by proving:

Proposition 19 *Let f be dominantly integrable, and let $0 < a < 1$. Then f is Riemann integrable on $[a, 1]$.*

Proof Fix $0 < a < 1$. It suffices to prove that, given $\varepsilon > 0$, there is a partition $\mathcal{P}$ of $[a, 1]$ such that

$$U(f, \mathcal{P}) - L(f, \mathcal{P}) \le \varepsilon. \tag{96}$$

Let η, δ correspond to the choice $\varepsilon/4$ in (95). There are then two cases: $0 < \eta \le a$, or $a < \eta$. If $a < \eta$, pick points $a = x_0 < x_1 < \ldots < x_n = 1$ such that $x_{k-1} > (1 - \delta)x_k$, and tags $c_k, c_k' \in I_k = [x_{k-1}, x_k]$, for $k = 1, \ldots, n$, so that, if $C = \{c_1, \ldots, c_n\}$ and $C' = \{c_1', \ldots, c_n'\}$, then $U(f, \mathcal{P}) \le S(f, \mathcal{P}, C) + \varepsilon/4$, and $S(f, \mathcal{P}, C') \le L(f, \mathcal{P}) + \varepsilon/4$. Then, for the partition $\mathcal{P} = \{I_k\}$ of $[a, 1]$, we have

$$U(f, \mathcal{P}) - L(f, \mathcal{P}) \le \left| S(f, \mathcal{P}, C) - S(f, \mathcal{P}, C') \right| + \varepsilon/2, \tag{97}$$

where the first term on the right-hand side of (97) is bounded by

$$\left| S(f, \mathcal{P}, C) - D \right| + \left| S(f, \mathcal{P}, C') - D \right| \le \varepsilon/2,$$

which, combined with (97), gives (96) in this case.

On the other hand, if $0 < \eta \le a$, pick $0 < x_0 < \eta$, and $y_0 = x_0 < y_1 < \ldots < y_m = a$, tags $d_h \in J_h = [y_{h-1}, y_h]$, such that $y_{h-1} > (1 - \delta)y_h$ for $h = 1, \ldots, m$. Then $\mathcal{Q} = \{J_h\} \cup \{I_k\}$ is a partition of $[x_0, 1]$ and is tagged. Let $C_1 = \{d_1, \ldots, d_m\} \cup C$, and $C_1' = \{d_1, \ldots, d_m\} \cup C'$; by (95), we have that the first term on the right-hand side in (97) is bounded by

$$\left| S(f, \mathcal{Q}, C_1) - D \right| + \left| S(f, \mathcal{Q}, C_1') - D \right| \le \varepsilon/2.$$

Hence, (96) also holds in this case, and we have finished. $\square$

In working with dominated integration, different approaches are important in applications, e.g., oscillations. We say that a function f defined on $(0, 1]$ satisfies the Riemann condition for the dominated integral, or *Property (R)*, provided that the following two conditions hold: (i) f is bounded on each closed subinterval of $(0, 1]$; and (ii) For every $\varepsilon > 0$, there exists $0 < \delta < 1$, such that

$$\sum_{k=1}^{n} \mathrm{osc}\,(f, I_k)\,|I_k| \leq \varepsilon, \tag{98}$$

whenever $0 < x_0 < x_1 < \ldots < x_n = 1$, $I_k = [x_{k-1}, x_k]$, and $x_k > (1 - \delta)x_{k-1}$ for $k = 1, \ldots, n$.

Note that if f satisfies Property (R), since

$$\big||f(x)| - |f(y)|\big| \leq |f(x) - f(y)|, \quad x, y \in (0, 1],$$

it readily follows that for a closed subinterval $J \subset (0, 1]$, $\mathrm{osc}\,(|f|, J) \leq \mathrm{osc}\,(f, J)$, and, therefore, also $|f|$ satisfies Property (R). Moreover, if F is defined on $(0, 1]$ by

$$F(x) = \sup_{x \leq t \leq 1} |f(t)|, \tag{99}$$

F is a monotone nonincreasing function such that $F(x) \geq |f(x)|$ throughout I, and also verifies Property (R). To see this, let $J = [a, b]$, $0 < a < b < 1$. Since F is nonincreasing, $F(a) - F(b) \geq 0$, and it suffices to consider when $F(a) - F(b) > 0$, and then we may restrict the sup in the definition of $F(a)$ to the set $a \leq t \leq b$. Then, for any $x_1, x_2 \in J$, we have

$$|F(x_1) - F(x_2)| \leq F(a) - F(b) \leq F(a) - |f(b)|$$

$$\leq \sup_{a \leq x \leq b} (|f(x)| - |f(b)|) \leq \sup_{a \leq x, x' \leq b} \big||f(x)| - |f(x')|\big|$$

$$= \mathrm{osc}\,(|f|, J) \leq \mathrm{osc}\,(f, J),$$

and, consequently,

$$\mathrm{osc}\,(F, J) \leq \mathrm{osc}\,(f, J). \tag{100}$$

Let f be a function defined on $(0, 1]$, and let $0 < \delta < 1$ be fixed. Suppose there is a real number I with the following property: For every $\varepsilon > 0$, there exists $0 < \eta < 1$ such that

$$\left|\sum_{k=1}^{n} f(c_k)(x_k - x_{k-1}) - I\right| \leq \varepsilon \tag{101}$$

whenever $0 = x_0 < x_1 < \ldots < x_n = 1$, the tags c_k satisfy max $(x_{k-1}, \delta x_k) \leq c_k \leq x_k$, and $x_k - x_{k-1} \leq \eta$ for $k = 1, \ldots, n$.

It is clear that if I exists, it is unique. We expect that such an f has a convergent improper integral, and the first step in verifying this is:

Proposition 20 *Suppose that f satisfies (101). Then f is Riemann integrable on $[a, 1]$ for any $0 < a < 1$.*

Proof The proof follows along the lines to that of Proposition 19, and we shall be brief. Given $\varepsilon > 0$, it suffices to produce a partition $\mathcal{P}$ of $[a, 1]$ such that $U(f, \mathcal{P}) - L(f, \mathcal{P}) \leq \varepsilon$. With $\delta < 1$ fixed, let η correspond to the choice $\varepsilon/4$ in (95); we may assume that $\eta < (1 - \delta)$. Consider the sequence $(1 - \eta)^n \to 0$, and let N be the largest integer such that $a < (1 - \eta)^N$. Consider the intervals $I_0 = [a, (1 - \eta)^N]$, $I_1 = [(1-\eta)^N, (1-\eta)^{N-1}], \ldots, I_{N-1} = [(1-\eta)^2, (1-\eta)], I_N = [(1-\eta), 1]$. Then, it readily follows that $|I_k| \leq \eta$ for $0 \leq k \leq N$, and with $a = x_{-1}$, if $I_k = [x_{k-1}, x_k]$, then max$(x_{k-1}, \delta x_k) = x_{k-1}$ for all k. Now, $\mathcal{P} = \{I_0, \ldots, I_N\}$ is a partition of $[a, 1]$, and for any set of tags C, C' by the definition above, it follows that since max$(x_{k-1}, \delta x_k) = x_{k-1}$, we may choose freely $c_k \in I_k$. Then we pick tags C, C' so that the Riemann sums $S(f, \mathcal{P}, C)$ approximate $U(f, \mathcal{P})$, and the Riemann sums $S(f, \mathcal{P}, C')$ approximate $L(f, \mathcal{P})$. Then $|S(f, \mathcal{P}, C) - S(f, \mathcal{P}, C')| \leq \varepsilon/2$, which gives that $U(f, \mathcal{P}) - L(f, \mathcal{P}) \leq \varepsilon$, and the assertion follows from the integrability criteria.   □

We then have:

Theorem 25 *Let f be defined on $(0, 1]$. Then the following are equivalent:*

(i) *f satisfies (101).*
(ii) *f is dominantly integrable on I.*
(iii) *f satisfies Property (R).*
(iv) *If F is defined as in (99), the improper integral of F converges.*

Moreover, in this case, the improper integral of f converges, and it equals D, where D is the real number in (95), and I, where I is the real number in (101).

Proof (i) implies (ii). Since f is Riemann integrable on $[a, 1]$ for $0 < a < 1$, it is bounded there. We claim that $xf(x) \to 0$ as $x \to 0^+$. Suppose this is not the case. Let η correspond to $\varepsilon = 1$ in (101), and observe there exist $\varepsilon > 0$ and a sequence $\{s_k\}$ with $\eta > s_1 > (1/2)s_1 > s_2 > \ldots > s_{k-1} > (1/2)s_{k-1} > s_k > \ldots > 0$, such that $|s_k f(s_k)| \geq \varepsilon$ for all k. By working with $-f$ and passing to a subsequence if necessary, we may assume that $s_k f(s_k) \geq \varepsilon$ for such a sequence. Finally, pick an integer N_1 such that $2^{-N_1} \leq \eta$, for integer $N > N_1$ let $k(N, N_1) = N - N_1 + 2^{N_1}$,, and for $1 \leq k \leq k(N, N_1)$, define the x_k^N as follows.

$$x_k^N = \begin{cases} 0, & k = 0, \\ s_{N-k}, & 1 \leq k \leq N - N_1, \\ s_{N_1} + (j - N + N_1)2^{-N_1}(1 - s_{N_1}), & N - N_1 + 1 \leq k \leq k(N, N_1). \end{cases}$$

Then, $x_k^N - x_{k-1}^N \leq \eta$ for all k, and for appropriate tags c_k, by (96),

$$\left| I - \sum_{k=1}^{2^{N_1}} f(c_k^N)\left(x_k^N - x_{k-1}^N\right) \right| \leq 1. \tag{102}$$

Now, we break up the sum in (102) into two terms, one summing up to $N - N_1$ and the other from $(N - N_1) + 1$ to 2^{N_1}, and observe that

$$\sum_{k=1}^{N-N_1} f(c_k^N)\left(x_k^N - x_{k-1}^N\right) \geq \frac{1}{2} \sum_{j=1}^{N-N_1} f(x_k^N)x_k^N \geq \frac{1}{2}(N - N_1)\,\varepsilon, \tag{103}$$

and, with M a bound for f on $[s_{N_1}, 1]$,

$$\left| \sum_{k=N-N_1+1}^{k(N,N_1)} f(c_k^N)\left(x_k^N - x_{k-1}^N\right) \right|$$

$$= 2^{-N_1}(1 - s_{N_1}) \left| \sum_{k=1}^{2^{N_1}} f(s_{N_1} + (2^{N_1} + 1 - k)2^{-N_1}(1 - s_{N_1})) \right|, \tag{104}$$

the above is $\leq M$. Hence, combining (102), (103), and (104), we have

$$1 \geq \left| I - \sum_{k=1}^{2^{N_1}} f(c_k^N)\left(x_k^N - x_{k-1}^N\right) \right| \geq \frac{1}{2}(N - N_1)\,\varepsilon - (M + |I|),$$

which is not true for N is sufficiently large.

Now we are ready for the proof. Given $\varepsilon_1 > 0$, we must find numbers D and $0 < \delta_1, \eta_1 < 1$ such that $\left| D - \sum_{k=1}^n f(c_k)(x_k - x_{k-1}) \right| \leq \varepsilon_1$ whenever $0 < x_0 < x_1 < \ldots < x_n = 1$, $x_0 < \eta_1$, $c_k \in [x_{k-1}; x_k]$, and $x_k - x_{k-1} \leq \delta_1 x_k$.

Let $0 < \delta < 1$ be given in (101), and let η be the value corresponding to $\varepsilon_1/2$ in (101). With τ such that $|xf(x)| \leq \varepsilon_1/2$ for $0 < x < \tau$, the values of η_1, δ_1 are defined by $\eta_1 = \min(\tau, \eta)$ and $\delta_1 = \min(\eta, 1 - \delta)$. Note that with these choices, since $x_0 \leq \eta_1 \leq \tau$, we have $|x_0 f(x_0)| \leq \varepsilon_1/2$. And, since $x_k - x_{k-1} \leq \delta_1 x_k \leq (1 - \delta)x_k$, then $\max(x_{k-1}, \delta x_k) = x_{k-1}$, and so c_k is an arbitrary point in $I_k = [\max(x_{k-1}, \delta x_k), x_k] = [x_{k-1}, x_k]$ for $1 \leq k \leq n$. Also, $x_k - x_{k-1} \leq \delta_1 x_k \leq \eta$ for $1 \leq k \leq n$. Then, with $c_{-1} = 0$ and $c_0 = x_0$, these observations give that $\sum_{k=0}^n f(c_k)(x_k - x_{k-1}) = f(c_0)x_0 + \sum_{k=1}^n f(c_k)(x_k - x_{k-1})$ is an admissible sum for (101), and, consequently,

$$\left| \sum_{k=1}^{n} f(c_k)(x_k - x_{k-1}) - I \right|$$

$$\leq \left| \sum_{k=0}^{n} f(c_k)(x_k - x_{k-1}) - I \right| + |f(x_0)x_0| \leq \varepsilon_1/2 + \varepsilon_1/2 = \varepsilon_1.$$

Hence, (95) holds with $D = I$.

(ii) implies (iii). Since f is Riemann integrable on closed subintervals of $(0, 1]$, it is bounded on each such subinterval. Given $\varepsilon > 0$, choose $0 < \delta, \eta < 1$ corresponding to $\varepsilon/4$ in (95). Let $0 < x_0 < x_1 < \ldots < x_n = 1$ such that $x_{k-1} > (1 - \delta) x_k$, and tags $c_k, c'_k \in I_k = [x_{k-1}, x_k]$ for $k = 1, \ldots, n$. Let $C = \{c_k\}$, $C' = \{c'_k\}$.

There are then two cases: $x_0 < \eta$, or $x_0 \geq \eta$. If $x_0 < \eta$, by (95),

$$\left| \sum_{k=1}^{n} \left(f(c_k) - f(c'_k) \right) |I_k| \right|$$

$$\leq \left| \sum_{k=1}^{n} f(c_k)|I_k| - D \right| + \left| \sum_{k=1}^{n} f(c'_k)|I_k| - D \right| \leq \varepsilon/2.$$

Now, pick c_k, c'_k such that osc $(f, I_k) \leq \left(f(c_k) - f(c'_k) \right) + \varepsilon/2n$ and note that by (101) above

$$\sum_{k=1}^{n} \text{osc}\,(f, I_k)\,|I_k| \leq \sum_{k=1}^{n} \left(f(c_k) - f(c'_k) \right)|I_k| + \varepsilon/2 \leq \varepsilon,$$

which gives (98) in this case.

On the other hand, if $x_0 \geq \eta$, pick $0 < y_0 < \eta$, $y_0 < y_1 < \ldots < y_m = x_0$, such that $y_{h-1} > (1 - \delta) y_h$, and tags $d_h \in J_h = [y_{h-1}, y_h]$, for $h = 1, \ldots, m$. Let $D = \{d_1, \ldots, d_m\}$. Then $Q = \{J_h\}$ is a partition of $[x_0, a]$, and by (95), it follows that

$$\sum_{k=1}^{n} \text{osc}\,(f, I_k)\,|I_k| \leq \sum_{k=1}^{n} \text{osc}\,(f, I_k)|I_k| + \sum_{h=1}^{m} \text{osc}\,(f, J_h)|J_h| \leq \varepsilon,$$

which gives (98) in this case, and we have finished.

(iii) implies (iv). Let δ be the value corresponding to $\varepsilon = 1$ in Property (R), and set $\eta = (1 - \delta/2) > (1 - \delta)$. We pick points $0 < x_0 = \eta^n = (1 - \delta/2)^n < x_1 = \eta^{n-1} < \ldots < x_{n-1} = \eta^1 < x_n = 1$ that satisfy the criteria in Property (R), and consider the intervals $I_k = [\eta^k, \eta^{k-1}]$ each of length $(1 - \eta)\eta^{k-1}$. Then, by (100) and Property (R),

$$\sum_{k=1}^{n} \left(F(\eta^k) - F(\eta^{k-1}) \right) |I^k| \le \sum_{k=1}^{n} \text{osc}\,(f, I_k)\,|I_k| \le 1.$$

Next, since by (99), F is positive and monotone decreasing, F is Riemann integrable on each $[a, 1]$, $0 < a < 1$. Since $F \ge 0$, it is improper Riemann integrable on $(0, 1]$, if $\int_{[\eta^n, 1]} F$ is uniformly bounded for n.

Now, since

$$F(\eta^{k-1})\,|I_k| \le \int_{[\eta^k, \eta^{k-1}]} F \le F(\eta^k)\,|I_k|, \quad k = 1, \ldots,$$

it follows that

$$\int_{[\eta^n, 1]} F \le \sum_{k=1}^{n} F(\eta^k)\,|I_k| = \sum_{k=1}^{n} F(\eta^k)\left(\eta^{k-1} - \eta^k\right)$$

$$= F(\eta) + \sum_{k=2}^{n} F(\eta^k)\,\eta^{k-1} - \sum_{k=2}^{n} F(\eta^{k-1})\,\eta^{k-1} - F(\eta^n)\eta^n$$

$$\le F(\eta) + \frac{1}{(1-\eta)} \sum_{k=2}^{n} \left(F(\eta^k) - F(\eta^{k-1}) \right) |I_k| \le F(\eta) + \frac{1}{(1-\eta)},$$

and the integral is bounded uniformly in n. This does it.

(iv) implies (i). Since $|f(x)| \le F(x)$ throughout $(0, 1]$, f is absolutely improper integrable. Let $\varepsilon > 0$ be given. Then pick γ such that

$$\left| \int_{[0^+, 1]} f - \int_{[\gamma, 1]} f \right|, \quad \frac{1}{\delta} \int_{[0^+, \gamma]} F \le \varepsilon/8,$$

and pick η such that for all partitions of $[\gamma, 1]$ with mesh $\|\mathcal{P}\| \le \eta$,

$$\left| S(f, \mathcal{P}, C) - \int_{[\gamma, 1]} f \right| \le \varepsilon/4.$$

Since η can be arbitrarily small, we may assume that $\eta M \le \varepsilon/8$, where M is a bound for f in $[\gamma, 1]$.

This is the value of η in definition (101). So, suppose that $0 < x_0 < x_1 < \ldots < x_n = 1$, satisfy $\max(x_{k-1}, \delta x_k) \le c_k \le x_k$, and $|I_k| = x_k - x_{k-1} \le \eta$, all k.

We will verify that

$$\left| \int_{[0^+,1]} f - \sum_{k=1}^{n} f(c_k)(x_x - x_{k-1}) \right| \le \varepsilon. \tag{105}$$

With N the largest integer such that $x_N < \gamma$, write

$$\sum_{k=1}^{n} f(c_k)(x_x - x_{k-1}) = \sum_{k=1}^{N} f(c_k)(x_x - x_{k-1}) + \sum_{k=N+1}^{n} f(c_k)(x_x - x_{k-1}).$$

Then, since $c_k \ge \delta x_k$,

$$\left| \sum_{k=1}^{N} f(c_k)(x_k - x_{k-1}) \right| \le \sum_{k=1}^{N} F(\delta x_k)(x_k - x_{k-1})$$

$$\le \int_{[0^+,\gamma]} F(\delta x)\, dx \le \frac{1}{\delta} \int_{[0^+,\delta\gamma]} F \le \varepsilon/4.$$

Let now $I'_N = [\gamma, x_{N+1}]$, and let $\mathcal{P} = \{I'_N, I_{N+1}, \ldots, I_n\}$. Then $\mathcal{P}$ is a partition of $[\gamma, 1]$, and $S(f, \mathcal{P}, C) = f(d_N)|[x_{N+1} - \gamma]| + \sum_{k=N+1}^{n}$ is a Riemann sum that satisfies the above conditions. Note that $|f(d_N)|\,|[x_{N+1} - \gamma]| \le M\eta \le \varepsilon/4$.

First, estimate (100) by

$$\left| \int_{[0^+,1]} f - \sum_{k=1}^{n} f(c_k)(x_x - x_{k-1}) \right|$$

$$\le \left| \int_{[0^+,1]} f - \int_{[\gamma,1]} f \right| + \left| \int_{[\gamma,1]} f - \sum_{k=1}^{n} f(c_k)(x_x - x_{k-1}) \right|,$$

where the first summand is bounded by $\varepsilon/8$.

As for the second summand, since

$$\sum_{k=1}^{n} f(c_k)(x_k - x_{k-1}) = \sum_{k=1}^{N} f(c_k)(x_k - x_{k-1} + S(f, \mathcal{P}, C) - f(d_N)|[x_{N+1} - \gamma]|,$$

it can be estimated by

$$\left| \int_{[\gamma,1]} f - S(f, \mathcal{P}, C) \right| + \left| \sum_{k=1}^{N} f(c_k)(x_k - x_{k-1}) \right| + \left| f(d_N)|[x_{N+1} - \gamma]| \right|,$$

where first term estimated by integration on $[\gamma, 1]$, second estimated above by improper integral of F, and third term $\le \varepsilon/4$. This completes the proof.   □

From Theorem 25, we highlight the fact that if the dominated integral of f exists, then there is a monotone nonincreasing function F on $(0, 1]$ such that: (i) $|f| \leq F$ throughout $(0, 1]$ and (ii) the improper Riemann integral of F converges. For functions that enjoy these properties, we have:

Theorem 26 *Let f be a function defined on $(0, 1]$, which is dominantly integrable on I. Then the improper integral of $|f|$ converges and conclusions (iii) and (iv) of Theorem 22 are valid for f.*

Proof First, if f is dominantly integrable on I, f is improper Riemann integrable on I. And, if F is given by (99), by (iv) of Theorem 25, the improper integral of F converges, and so does the improper integral of $|f|$. Also, since F is monotone, conclusions (iii) and (iv) of Theorem 22 hold for F. Moreover, since $|f(x)| \leq F(x)$ throughout $(0, 1]$, by Theorem 23, if Π is a class of partitions of $[0, 1]$ containing partitions $\mathcal{P}$ with arbitrarily small mesh $\|\mathcal{P}\|$ such that $\lim_{\|\mathcal{P}\| \to 0} S_\Pi(F, \mathcal{P}, C) = \int_{[0^+, 1]} F$, then for the same family Π of partitions $\mathcal{P}$, $\lim_{\|\mathcal{P}\| \to 0} S_\Pi(f, \mathcal{P}, C) = \int_{[0^+, 1]} f$. Since clearly the partitions identified in (iii) and (iv) of Theorem 22 satisfy the assumptions of Theorem 23, the proof is finished.   □

We will now consider the evaluation of an improper integral by means of quadrature rules. Because of the nature of improper integrals, we will restrict ourselves to the open type rules that do not include the left endpoint. We will refer back to the previously established fact that for Riemann integrable functions f on $I = [0, 1]$, the compound rules converge to $\int_I f$.

The following circumstance is appropriate when dealing with dominantly integrable functions. Fix $0 < \delta < 1$, and consider a family of partitions $\{\mathcal{P}_n\}$ with $\lim_n \|\mathcal{P}_n\| = 0$, as those considered in (95). Thus, each partition $\mathcal{P}_n = \{J_k^n\}$ consists of intervals $J_k^n = [x_{k-1}^n, x_k^n]$, $1 \leq k \leq m_n$, and tags $c_1^n < \ldots < c_{m_n}^n$, such that $c_k^n \in J_k^n$ for all k, and

$$\max\left(x_{k-1}^n, \delta x_k^n\right) \leq c_k^n \leq x_k^n, \quad k = 1, 2, \ldots, m_n. \tag{106}$$

We then say that the sequence $\{\Phi_n\}$ of functionals on the set of functions h defined on $(0, 1]$ is a *Q-sequence* (where Q stands for quadrature) if each Φ_n is given by

$$\Phi_n(h) = \sum_{k=1}^{m_n} w_k^n h(c_k^n)\left(x_k^n - x_{k-1}^n\right), \quad n = 1, 2, \ldots, \tag{107}$$

with the property that for every function f that is Riemann integrable on I,

$$\lim_n \Phi_n(f) = \int_I f. \tag{108}$$

Now, applying (107) and (108) to $h = \chi_I$, it readily follows that a necessary condition in this case is that

$$\lim_n \sum_{k=1}^{m_n} w_k^n \left(x_k^n - x_{k-1}^n\right) = 1,$$

which we will assume. We will also assume that $|w_k^n| \leq M$, for $k = 1, \ldots m_n$, $n = 1, 2, \ldots$ We then have:

Theorem 27 *Let f be a function defined on $(0, 1]$. Then, f is dominantly integrable iff each Q-sequence $\{\Phi_n\}$ converges, and in this case $\lim_n \Phi_n(f) = \int_{[0^+,1]} f$.*

Proof Sufficiency follows readily from Theorem 25. Indeed, if f is not dominantly integrable, for every $0 < \delta < 1$, there is a Q-sequence $\{\Phi_n\}$ (each $\Phi_n(f)$ being a Riemann sum of f) such that $\{\Phi_n(f)\}$ does not converge.

As for necessity, note that by Theorem 25 the improper integral of f converges. So, given $\varepsilon > 0$, we will verify that $|\int_{[0^+,1]} f - \Phi_n(f)| \leq \varepsilon$ provided that n is sufficiently large. First, since the improper integral of f converges, for $\eta_1 < 1$ sufficiently small, it follows that

$$\left|\int_{[0^+,\delta\eta_1]} f\right| < \varepsilon/3. \tag{109}$$

Also for $\eta_2 < 1$, consider

$$\sum_{k=1}^{m_n} w_k^n \chi_{[0,\delta\eta_2]}(c_k^n) f(c_k^n)\left(x_k^n - x_{k-1}^n\right),$$

and observe that the sum extends over those k such that $c_k^n \leq \delta\eta_2$, which together with (106) implies that $\delta x_k^n \leq c_k^n \leq \delta\eta_2$, or $x_k^n \leq \eta_2$.

Now, since by (iv) of Theorem 25, F is decreasing and has an improperly convergent integral, for $\eta_2 < 1$ sufficiently small, it follows that the above sum can be estimated by

$$\sum_{k=1}^{m_n} \left|\chi_{[0,\delta\eta_2]}(c_k^n) w_k^n f(c_k^n)\left(x_k^n - x_{k-1}^n\right)\right| \leq M \sum_{k=1}^{m_n} \chi_{\{x_k^n \leq \eta_2\}}(k) F(c_k^n)\left(x_k^n - x_{k-1}^n\right)$$

$$\leq M \sum_{k=1}^{m_n} \chi_{\{x_k^n \leq \eta_2\}}(k) F(\delta x_k^n)\left(x_k^n - x_{k-1}^n\right) \leq M \int_{[0^+,\eta_2]} F(\delta x)\, dx$$

$$= \frac{M}{\delta} \int_{[0^+,\delta\eta_2]} F \leq \varepsilon/3, \tag{110}$$

for η_2 sufficiently small, independently of n. If we pick now $\eta = \min(\eta_1, \eta_2)$, (109) and (110) hold simultaneously for η, independently of n.

Let now the Q-sequence $\{\Phi_n(h)\}$ be defined by (107), and note that since $\chi_{[\delta\eta,1]}(x) + (1 - \chi_{[\delta\eta,1]}(x)) = 1$ for all x, $\Phi_n(h)$ can be expressed as

$$\Phi_n(f) = \sum_{k=1}^{m_n} w_k^n \left(1 - \chi_{[\delta\eta,1]}(c_k^n)\right) f(c_k^n) \left(x_k^n - x_{k-1}^n\right)$$

$$+ \sum_{k=1}^{m_n} w_k^n \, \chi_{[\delta\eta,1]}(c_k^n) \, f(c_k^n) \left(x_k^n - x_{k-1}^n\right).$$

Now, since $\int_{[0^+,1]} f = \int_{[0^+,\delta\eta]} f + \int_{[\delta\eta,1]} f$, with the above representation for $\Phi_n(f)$, it follows that

$$\left| \int_{[0^+,1]} f - \Phi_n(f) \right| \leq \left| \int_{[0,\delta\eta]} f \right|$$

$$+ \sum_{k=1}^{m_n} \chi_{[0,\delta\eta]}(c_k^n) \left| w_k^n f(c_k^n) \left(x_k^n - x_{k-1}^n\right) \right|$$

$$+ \left| \int_{[\delta\eta,1]} f - \sum_{k=1}^{m_n} w_k^n \chi_{[\delta\eta,1]}(c_k^n) f(c_k^n) \left(x_k^n - x_{k-1}^n\right) \right|,$$

where the sum of the first plus the second terms above is estimated by $\varepsilon/3 + \varepsilon/3 = 2\varepsilon/3$, independently of n. As for the third term, by (108), it can be made $\leq \varepsilon/3$ provided n is large enough, and the proof is finished. $\qquad\Box$

An interesting application of this result is to compound quadrature rules. Suppose that $\{R_n(f)\}$ is a sequence of open compound rules on I not involving $f(0)$, integrating χ_I exactly, namely,

$$R_n(f) = \frac{1}{n} \sum_{k=1}^{n} \left(\sum_{r=1}^{m} w_r \, f\left(\frac{(k-1)}{n} + \frac{x_r}{n}\right)\right), \quad n = 1, 2, \ldots,$$

where $w_1, \ldots, w_m$ are real numbers such that $\sum_{r=1}^{m} w_r = 1$, and $0 < x_1 < \ldots < x_m = 1$.

Observe that for $n = 2, 3, \ldots$, the numbers

$$\frac{(k-1)}{n} + \frac{x_r}{n}, \quad k = 1, 2, \ldots, n, \ r = 1, 2, \ldots, m$$

are all distinct, and arrange them as a strictly monotone increasing sequence τ_j^n, where $1 \leq j \leq mn$, and set for each n,

$$
t_j^n = \begin{cases} 0, & j = 0, \\ \dfrac{(\tau_j^n + \tau_{j+1}^n)}{2}, & j = 1, 2, \ldots, nm - 1, \\ 1, & j = nm. \end{cases}
$$

Observe that the above definition of τ_j^n associates with each $n > 1$ and $j = 1, 2, \ldots, nm$, a unique r, $1 \leq r \leq m$. Then, given such n and j, use the corresponding r to define

$$
c_j^n = \frac{w_r}{n \, (t_j^n - t_{j-1}^n)}.
$$

From the definition of the t_j^n, it follows that there exists a constant $M_1^{-1} > 0$ (independent of j and n) such that $t_j^n - t_{j-1}^n > (M_1 n)^{-1}$ for every j, n; thus, each $|c_j^n| < M$, with M being a constant. It is also verified that, for $n = 2, 3, \ldots$ and $j = 1, 2, \ldots, nm$,

$$
\tau_j^n / t_j^n \geq \delta = 2 \left(1 + \max_{1 \leq r \leq m} (x_{r+1}/x_r)\right)^{-1},
$$

where $x_{m+1} = 1 + x_1$.

Recall that for every Riemann integrable function f on I, $R_n(f)$ converges to $\int_I f$. Hence, by Theorem 27, $\lim_n R_n(f) = \int_{[0^+, 1]} f$, for dominantly integrable functions f, [73].

6.3   Improper Integrals of the Second Type

We will now discuss functions of the second improper type. A function f defined on $[0, \infty)$ and Riemann integrable on $[0, r)$ for all r is said to have an improper Riemann integral on $[0, \infty)$ if the limit

$$
\lim_{r \to \infty} \int_{[0,r]} f = \int_{[0,\infty)} f
$$

exists. Such functions are not necessarily Lebesgue integrable on $[0, \infty)$, as the example $f(x) = \sum_{k=1}^{\infty} (-1)^k (1/k) \chi_{[k,k+1)}(x)$ shows.

Keeping in mind the interplay between the monotonicity of the function and the selection of partitions involved, our purpose is to define the improper integral of f by means of a single limit involving an infinite sum. Suppose that f is nonnegative monotone decreasing, and let $0 = x_0 < x_1 < x_2 < \ldots,$ be the points of division of the intervals in the partition. Then it is clear that

$$\sum_{n=0}^{\infty} (x_{n+1} - x_n) f(x_n) \geq \int_{[0,\infty)} f \geq \sum_{n=0}^{\infty} (x_{n+1} - x_n) f(x_{n+1}),$$

in so far as these inequalities have a meaning, that is to say, the convergence of the integral ensures that of the second series and the second part of the inequality, and the convergence of the first series ensures both inequalities. It is easy to see, by an example, that the convergence of the integral does not ensure that of the first series above. Take, e.g., $f(x) = 1/(1 + x)^2$ and $x_n = 2^{2^n} h$, [12].

However, in the most interesting case, the set of partition intervals of $[0, \infty)$ is that divided by the points $0 < h < 2h < \ldots,$ where h is a parameter that tends to 0, i.e., $x_n = nh$. Then the monotonic character f and the convergence of the integral, or of either series, are sufficient to ensure that all three converge and that the limit of either series as $h \to 0^+$ is equal to the value of the integral. For, if $h(f(h) + f(2h) + \cdots)$ is convergent, so also is $h(f(0) + f(h) + f(2h) + \cdots)$, and the difference of their sums, i.e., $hf(0)$, tends to zero as $h \to 0$. Thus

$$\lim_{h \to 0^+} h \sum_{n=1}^{\infty} f(nh) = \int_{[0,\infty)} f.$$

Bromwich and Hardy study the existence, and evaluation, of improper integrals of the second type by means of sums of the form $h \sum_{k=1}^{\infty} f(hk)$ for positive values of h. Now, in this setting, monotonicity is essential; the fact that f is positive and continuous and that the improper integral of f is convergent is not enough to ensure the convergence of such sums for all positive values of h. To see this, let $f(x) = 1$ whenever $x = 2^n$. Then, if $h = 2^{-n}$, the series includes the terms $f(1), f(2), f(4), \ldots,$ and therefore diverges. But, graphically or otherwise, it is easy to define a continuous function f that has the values prescribed above, and yet its improper integral converges to any assigned value.

The situation is different, however if f decreases steadily as x increases to ∞, and an important ingredient in the proofs is Abel's Lemma. Specifically:

Abel's Lemma Let $\{u_k\}, \{v_k\}$ be sequences, and let $U_k = \sum_{h=1}^{k} u_h$. Then,

$$\sum_{k=1}^{n} u_k v_k = \sum_{k=1}^{n-1} U_k (v_k - v_{k+1}) + U_n v_n.$$

Moreover, if $v_k \geq 0$, and the sequence is decreasing, then

$$\left| \sum_{k=1}^{n} u_k v_k \right| \leq v_1 \max |U_k|.$$

We are now ready to prove the convergence results:

Theorem 28 *Let f be a function defined for $x \geq 0$ that is Riemann integrable on $[0, r]$ for all r and that satisfies the following property: there are constants K and M so that*

$$\left| \int_{[\xi, \xi']} f \right| \leq K, \quad \text{for all } M < \xi < \xi'. \tag{111}$$

Further suppose that φ is a nonincreasing function defined for $x \geq 0$ that is Riemann integrable on $[0, r]$ for all r, which tends to 0 as $x \to \infty$. Then, the improper Riemann integral $\int_{[0,\infty)} f\varphi$ converges.

Proof Since $f\varphi$ is integrable on $[0, r]$ for all r, to verify the conclusion, it suffices to prove that, given $\varepsilon > 0$, there exists M' such that

$$\left| \int_{[\xi, \xi']} f\varphi \right| \leq \varepsilon, \quad \text{for all } M' < \xi < \xi'.$$

Put $I = [\xi, \xi']$, and let $\{I_k^n\}$, $I_k^n = [x_{k,l}, x_{k,r}]$ be the partition of I in Π_e consisting of n equal sized intervals. Since f is integrable on I, f is bounded there by c, say, and, therefore, $\int_{I_k^n} |f| \leq c |I_k^n| = c |I|/n$, for all $1 \leq k \leq n$.

Write now

$$J = \int_I f\varphi = \sum_{k=1}^n \int_{I_k^n} f\varphi = \sum_{k=1}^n J_k,$$

say. Since $J_k - \varphi(x_{k,r}) \int_{I_k^n} f = \int_{I_k^n} f\left(\varphi - \varphi(x_{k,r})\right)$, by the monotonicity of φ, it follows that

$$\left| J_k - \varphi(x_{k,r}) \int_{I_k^n} f \right| \leq \left(\varphi(x_{k,l}) - \varphi(x_{k,r})\right) \int_{I_k^n} |f|.$$

Whence summing, since $x_{1,l} = \xi$, we have

$$\left| J - \sum_{k=1}^n \varphi(x_{k,r}) \int_{I_k^n} f \right| \leq \sum_{k=1}^n \left(\varphi(x_{k,l}) - \varphi(x_{k,r})\right) \int_{I_k^n} |f|$$

$$\leq \left(\max_k \int_{I_k^n} |f|\right) \left(\varphi(x_{1,l}) - \varphi(x_{n,r})\right) \leq \varphi(\xi) c |I|/n.$$

Also, by Abel's lemma,

$$\left| \sum_{k=1}^n \varphi(x_{k,r}) \int_{I_k^n} f \right| \leq \max_k \left| \sum_{j=1}^k \int_{I_j^n} f \right| \varphi(x_{1,r}),$$

where by (111) above since $\varphi(x_{1,r}) \leq \varphi(\xi)$, and with ξ' running over $\xi + (\xi' - \xi)k/n$,

$$\max_k \left| \sum_{j=1}^{k} \int_{I_j^n} f \right| = \max_k \left| \int_{[\xi, \xi + (\xi' - \xi)k/n]} f \right| \leq K,$$

gives that

$$\left| \sum_{k=1}^{n} \varphi(x_{k,r}) \int_{I_k^n} f \right| \leq K\varphi(\xi).$$

Therefore, for n sufficiently large, putting everything together,

$$|J| \leq \left| J \mp \sum_{k=1}^{n} \varphi(x_{k,r}) \int_{I_k^n} f \right| \leq \varphi(\xi)c|I|/n + K\,\varphi(\xi). \tag{112}$$

Finally, given $\varepsilon > 0$, pick M'' such that $\varphi(x) \leq \varepsilon/K$ for $M'' < x$, and let $M' = \max(M, M'')$. This gives the conclusion. $\qquad\qquad\square$

Our next result addresses the representation of the improper integral by means of a single limit. In this sense, we have:

Theorem 29 *Let f be a function defined for $x \geq 0$ that is Riemann integrable on $[0, r]$ for all r and that satisfies the following property: there are a small positive real α and a constant K, so that, for $0 < h < \alpha$ and all integers $n_1 < n_2$,*

$$h \left| \sum_{k=n_1}^{n_2} f(hk) \right| \leq K. \tag{113}$$

Further suppose that φ is a nonincreasing function defined for $x \geq 0$ that is Riemann integrable on $[0, r]$ for all r and that tends to 0 as $x \to \infty$. Then, the improper Riemann integral of $f\varphi$ converges, $\sum_{k=1}^{\infty} f(kh)\varphi(kh)$ converges for $h < \alpha$, and we have

$$\lim_{h \to 0^+} h \sum_{k=1}^{\infty} f(kh)\varphi(kh) = \int_{[0,\infty)} f\varphi. \tag{114}$$

Proof We begin by considering the integral. Since f is Riemann integrable on $[\xi, \xi']$, f is (U) integrable there, and, consequently,

$$\int_{[\xi, \xi']} f = (U) \int_{[\xi, \xi']} f = \lim_{h \to 0^+} h \sum_{k=[\xi/h]+1}^{[\xi'/h]} f(kh),$$

and so by (113), it follows that $\left| \int_{[\xi,\xi']} f \right| \leq K$. Then, by Theorem 28,

$$\left| \int_{[\xi,\xi']} f\varphi \right| \leq K\,\varphi(\xi), \quad \text{all } \xi' > \xi. \tag{115}$$

Now, given $\varepsilon > 0$, if we pick ξ large enough so that $K\varphi(\xi) \leq \varepsilon$, then from (115) it follows that the improper Riemann integral of $f\varphi$ converges. Furthermore, since (115) holds for all $\xi' > \xi$, we also have that

$$\left| \int_{[\xi,\infty)} f\varphi \right| \leq K\,\varphi(\xi). \tag{116}$$

As for the sum, for $h < \alpha$, by Abel's lemma, for any $n_1 < n_2$, we have

$$\left| h \sum_{k=n_1}^{n_2} f(kh)\,\varphi(kh) \right| \leq \max_{n_1 \leq m \leq n_2} \left| h \sum_{k=n_1}^{m} f(kh) \right| \varphi(n_1 h)$$

$$\leq K\,\varphi(n_1 h), \tag{117}$$

and, therefore,

$$\left| h \sum_{k=n_1}^{n_2} f(kh)\,\varphi(kh) \right| \leq K\varphi(\xi), \tag{118}$$

as long as $n_1 h \geq \xi$.

Now, with $h < \alpha$ fixed, given $\varepsilon > 0$, pick n_1 large enough so that $\varphi(n_1 h) \leq h\varepsilon/K$, and note that from (118) it follows that

$$\left| \sum_{k=n_1}^{n_2} f(kh)\,\varphi(kh) \right| \leq \varepsilon,$$

which implies that the sum on the left-hand side of (114) converges for each $h < \alpha$. Moreover, from (117), as long as $n_1 h \geq \xi$, it follows that

$$\left| h \sum_{k=n_1}^{\infty} f(kh)\,\varphi(kh) \right| \leq K\varphi(\xi). \tag{119}$$

Fix $\xi > 0$ now so that $\varphi(\xi) \leq \varepsilon/3K$. Then, since $f\varphi$ is Riemann integrable on $[0, \xi]$, and so (U) integrable there, we have

$$\int_{[0,\xi]} f\varphi = \lim_{h \to 0^+} h \sum_{k=1}^{[\xi/h]} f(kh)\varphi(kh).$$

Therefore, we can pick $0 < h < \alpha$ such that

$$\left| \int_{[0,\xi]} f\varphi - h \sum_{k=1}^{[\xi/h]} f(kh)\varphi(kh) \right| \leq \varepsilon/3. \tag{120}$$

With this h fixed, look at $[\xi/h] + 1$. Then,

$$h\left(\left[\frac{\xi}{h}\right] + 1\right) \geq h\left(\frac{\xi}{h}\right) = \xi,$$

and, consequently, $[\xi/h] + 1 = n_1$ in (119), and therefore,

$$\left| h \sum_{k=[\xi/h]+1}^{\infty} f(kh)\,\varphi(kh) \right| \leq K\varphi(\xi) \leq \varepsilon/3. \tag{121}$$

Hence, by the triangle inequality, combining the (120), (116), and (121),

$$\left| \int_{[0,\infty)} f\,\varphi - h \sum_{k=1}^{\infty} f(kh)\varphi(kh) \right|$$

$$\leq \left| \int_{[0,\xi]} f\,\varphi - h \sum_{k=1}^{[\xi/h]} f(kh)\varphi(kh) \right|$$

$$+ \left| \int_{[\xi,\infty)} f\varphi \right| + \left| h \sum_{k=[\xi/h]+1}^{\infty} f(kh)\,\varphi(kh) \right| \leq \varepsilon.$$

This completes the proof.   □

Interesting instances of Theorem 29 are obtained by letting $f(x) = \sin(x)$ or $f(x) = \cos(x)$. Now, the case when $\varphi(x) \to \infty$ as $x \to 0^+$ falls outside the scope of Theorem 29, but this difficulty is not essential [12]. Indeed, if all the conditions of Theorem 29 are satisfied except that $\varphi(x) \to \infty$ as $x \to 0+$ in such a way that $f\varphi$ is positive and monotone in a neighbourhood of the origin and $\int_{0+,a} f\varphi$ converges for some $a > 0$, then by (i) of Theorem 22, $f\varphi$ is (U) integrable on that neighbourhood of the origin, and (117) above holds. The proof proceeds then as before, and the conclusion of Theorem 29 holds.

This observation applies to the following setting [54]. Let

$$\psi(h) = \sum_{k=1}^{\infty} \frac{1}{k^{1+s}}\big(1 - \cos(kh)\big), \quad 0 < s < 2.$$

We are interested in evaluating

$$\lim_{h \to 0^+} \frac{\psi(h)}{h^s}, \quad \text{and,} \quad \lim_{h \to 0^+} \frac{\psi'(h)}{h^{s-1}}.$$

Observe that

$$\frac{1 - \cos x}{x^{1+s}} = 2\,\frac{\sin^2(x/2)}{x^{1+s}}$$

is positive and monotone near 0, and all the above conditions are met.
Then,

$$\lim_{h \to 0^+} \frac{\psi(h)}{h^s} = \lim_{h \to 0^+} h \sum_{k=1}^{\infty} \frac{1}{(kh)^{1+s}}\big(1 - \cos(kh)\big)$$

$$= \int_{[0,\infty)} \frac{1 - \cos x}{x^{1+s}}\,dx = \frac{\pi}{2s\,\Gamma(s)\sin(s\pi/2)}.$$

Similarly, since

$$\psi'(h) = \sum_{k=1}^{\infty} \frac{\sin(kh)}{k^s}, \qquad \frac{\psi'(h)}{h^{\alpha-1}} = h \sum_{k=1}^{\infty} \frac{\sin(kh)}{(kh)^s},$$

and, consequently,

$$\lim_{h \to 0^+} \frac{\psi'(h)}{h^{\alpha-1}} = \lim_{h \to 0^+} h \sum_{k=1}^{\infty} \frac{\sin(kh)}{(kh)^s}$$

$$= \int_{[0,\infty)} \frac{\sin(x)}{x^s}\,dx = \frac{\pi}{4\Gamma(s)\sin(s\pi/2)}.$$

We also have the following result, the proof of which being similar to that of Theorem 29 is left to the reader [12].

Theorem 30 *Let $f(x)$ be a function defined for $x \geq 0$ that is Riemann integrable on $[0, r]$ for all r and that satisfies the following property: there is a small positive real α, so that, for $0 < h < \alpha$, given $\varepsilon > 0$, there is an integer M such that for all integers $M \leq n_1 < n_2$,*

$$h\left|\sum_{k=n_1}^{n_2} f(hk)\right| \leq \varepsilon.$$

Then, f has a convergent improper Riemann integral given by

$$\lim_{h \to 0^+} h \sum_{k=1}^{\infty} f(kh) = \int_{[0,\infty)} f.$$

Furthermore, suppose that $\varphi(x)$ is a nonincreasing function defined for $x \geq 0$ that is Riemann integrable on $[0, r]$ for all r that tends to a limit $L \geq 0$ as $x \to \infty$. Then $f\varphi$ has a convergent improper Riemann integral given by

$$\lim_{h \to 0^+} h \sum_{k=1}^{\infty} f(kh)\varphi(kh) = \int_{[0,\infty)} f\,\varphi.$$

We will consider briefly the concept of *simple integral*, which informally corresponds to that of dominated integral for improper integrals of the first type. Let f be a real function defined on $[0, \infty)$. We say that f is *simply integrable* if there is a number I with the following property: For every $\varepsilon > 0$, there are numbers B and η such that

$$\left| \sum_{k=1}^{n} f(c_k)(x_k - x_{k-1}) - I \right| \leq \varepsilon,$$

provided that $b > B$, and $\mathcal{P} = \{J_k\}$ is a partition of $[0, b]$ with $\|\mathcal{P}\| < \eta$, and $J_k = [x_{k-1}, x_k]$, $1 \leq k \leq n$, with $x_0 = 0$, $x_n = b$, and the tags $c_k \in J_k$, $1 \leq k \leq n$.

The number I (which is unique) is the *simple integral* of f, [72]. Now, if f is simply integrable, f is improperly Riemann integrable and $I = \int_{[0,\infty)} f$.

The function $f(x) = (\sin x)/x$ is not simply integrable, while $g(x) = (\sin x^2)/x^2$ is. Informally, f is simply integrable if the Riemann sums of f associated with partitions $\mathcal{P}$ of $[0, b]$ approach a (finite) limit as long as $b \to \infty$ and $\|\mathcal{P}\| \to 0$ simultaneously.

There are equivalent formulations of the way we look at simple integrable functions. We say that a function f on $[0, \infty)$ satisfies the *infinite-Riemann-sum condition* on $[0, \infty)$ iff there is a number I with the following property: For each $\varepsilon > 0$, there is $\eta > 0$ such that if $0 = x_0 < x_1 < \ldots, x_k, x_k \to \infty, x_k - x_{k-1} < \eta$ and $x_{k-1} \leq c_k \leq x_k$ for $k = 1, 2, \ldots$ then the series $\sum_{k=1}^{\infty} f(c_k)(x_k - x_{k-1})$ converges and

$$\left| \sum_{k=1}^{\infty} f(c_k)(x_k - x_{k-1}) - I \right| \leq \varepsilon.$$

Then, f is simply integrable iff f satisfies the infinite-Riemann-sum condition on $[0, \infty)$, in which case $I = \int_{[0,\infty)} f$, [59]. And, in that paper, it is proved that a function f defined on $(0, 1]$ is dominantly integrable iff it satisfies the infinite-Riemann-sum condition on $(0, 1]$. In fact, it is the case that infinite sums are applicable to Riemann integrable functions as well [29]. Of course, there are similar characterizations with oscillations.

Chapter 7
Coda

It is natural, and of interest, to extend the results we have discussed to the context of Riemann–Stieltjes integration [100], with applications, for instance, to stochastic integrals [60]. In this closing chapter, we outline the extension of the modified Riemann sums, pattern and uniform integrals, distribution functions, and quadrature formulas. The proof of these results is built on our previous results.

We will begin by introducing the necessary definitions and notations. Fix a closed finite interval $I = [a, b] \subset \mathbb{R}$. We will henceforth assume that Ψ is an indefinite integral on I, i.e., there is a nonnegative Riemann integrable function ψ defined on I, such that

$$\Psi(x) = \Psi(a) + \int_{[a,x]} \psi, \quad x \in I. \tag{122}$$

Ψ is then a continuous, monotone increasing function defined on I; such functions have been characterized in [96].

Let Π be an admissible family of partitions of I. For a partition $\mathcal{P} = \{I_1, \ldots, I_m\}$ of I in Π, where $I_k = [x_{k,l}, x_{k,r}]$, $1 \le k \le m$, and a bounded function f on I, let $U(f, \Psi, \mathcal{P})$ and $L(f, \Psi, \mathcal{P})$ denote the upper and lower Riemann sums of f with respect to Φ on I along $\mathcal{P}$, i.e.,

$$U(f, \Psi, \mathcal{P}) = \sum_{k=1}^{m} \left(\sup_{I_k} f \right) \left(\Psi(x_{k,r}) - \Psi(x_{k,l}) \right),$$

and

$$L(f, \Psi, \mathcal{P}) = \sum_{k=1}^{m} \left(\inf_{I_k} f \right) \left(\Psi(x_{k,r}) - \Psi(x_{k,l}) \right),$$

respectively, and set

$$U_\Pi(f, \Psi) = \inf_{\mathcal{P} \in \Pi} U(f, \Psi, \mathcal{P}), \quad \text{and,} \quad L_\Pi(f, \Psi) = \sup_{\mathcal{P} \in \Pi} L(f, \Psi, \mathcal{P}).$$

We say that f is Π-*Riemann–Stieltjes integrable* with respect to Ψ on I if $U_\Pi(f, \Psi) = L_\Pi(f, \Psi)$, and in this case, the common value is denoted $\int_I f \, d\Psi$ and is called the Π-*Riemann–Stieltjes integral of f with respect to Ψ on I*. This integral has similar properties to the Π-Riemann integral, and the proofs will be omitted [100]. In particular, the sequence $\{S(f, \Psi, \mathcal{P}_n, C)\}$ of arbitrary Riemann sums of f with respect to Ψ on I corresponding to the tagged partitions $\{\mathcal{P}_n\} \subset \Pi_e$, consisting of the intervals $\mathcal{P}_n = \{I_1^n, \ldots, I_n^n\}$, with $I_k^n = [x_{k,l}^n, x_{k,r}^n]$, given by

$$S(f, \Psi, \mathcal{P}_n, C) = \sum_{k=1}^{n} f(c_k^n) \left(\Psi(x_{k,r}^n) - \Psi(x_{k,l}^n) \right), \quad c_k^n \in I_k^n,$$

which lie between $L(f, \Psi, \mathcal{P}_n)$ and $U(f, \Psi, \mathcal{P}_n)$, will also converge to the common limit above, i.e.,

$$\lim_n S(f, \Psi, \mathcal{P}_n, C) = \int_I f \, d\Psi. \tag{123}$$

Integrability can also be characterized in terms of the oscillation of a function [13, 46, 100]. A bounded function f is Riemann–Stieltjes integrable with respect to Ψ on I iff, given $\varepsilon > 0$, there is a partition $\mathcal{P} = \{I_k\}$ of I, which may depend on ε, such that

$$\sum_k \mathrm{osc}\,(f, I_k) \left(\Psi(x_{k,r}) - \Psi(x_{k,l}) \right) \le \varepsilon. \tag{124}$$

And, a sequential characterization holds, to wit, (124) is equivalent to the existence of a sequence of partitions $\{\mathcal{P}_n\}$ of I consisting of the intervals $\mathcal{P}_n = \{I_k^n\}$, with $I_k^n = [x_{k,l}^n, x_{k,r}^n]$, such that

$$\lim_n \sum_k \mathrm{osc}\,(f, I_k^n) \left(\Psi(x_{k,r}^n) - \Psi(x_{k,l}^n) \right) = 0. \tag{125}$$

Also, by the Theorem in Appendix I, f is Riemann–Stieltjes integrable with respect to Ψ on I iff $f\psi$ is Riemann integrable on I. Other properties, such as (5) above, also hold in this context with similar proofs.

We will now introduce the notion of modified Riemann sums . Let Φ be a set mapping defined on the subintervals J of I that assigns to each $J \subset I$ a subset

$J^1 = \Phi(J) \subset I$ with the following properties:

(i) $\chi_{\Phi(J)}$ is Riemann integrable, and so, the Ψ-length $|J^1|_\Psi$ of J^1 is well-defined as $|J^1|_\Psi = 0$ if $J^1 = \emptyset$, and $|J^1|_\Psi = \int_I \chi_{\Phi(J)} \psi$, otherwise.

(ii) There exists $\eta > 0$ such that $J^1 = \Phi(J) \subset J$, whenever $|J| < \eta$.

Given a partition $\mathcal{P}$ of I consisting of intervals $I_1, \ldots, I_m$, we will denote with $\mathcal{P}^1$ the collection of subsets of I consisting of $I_1^1 = \Phi(I_1), \ldots, I_m^1 = \Phi(I_m)$, say, and set

$$u(f, \Psi, \mathcal{P}^1) = \sum_{k=1}^m \left(\sup_{I_k^1} f\right) |I_k^1|_\Psi, \quad \text{and,} \quad l(f, \Psi, \mathcal{P}^1) = \sum_{k=1}^m \left(\inf_{I_k^1} f\right) |I_k^1|_\Psi.$$

We call these expressions the *modified upper* and *modified lower Riemann sums of* f *with respect to* Ψ *along* $\mathcal{P}$ *on* I, respectively, and set

$$u(f, \Psi) = \inf_{\mathcal{P}} u(f, \Psi, \mathcal{P}^1), \quad \text{and,} \quad l(f, \Psi) = \sup_{\mathcal{P}} l(f, \Psi, \mathcal{P}^1).$$

Note that $l(f, \Psi) \le u(f, \Psi)$. We say that *the modified Riemann sums of* f *with respect to* Ψ *on* I *converge* if $u(f, \Psi) = l(f, \Psi)$. In this case, the *(arbitrary) modified Riemann sums of* f *with respect to* Ψ *on* I given by

$$s(f, \Psi, \mathcal{P}^1, C^1) = \sum_{k=1}^m f(x_k^1) |I_k^1|_\Psi, \quad x_k^1 \in I_k^1, \tag{126}$$

which lie between $l(f, \Psi, \mathcal{P}^1)$ and $u(f, \Psi, \mathcal{P}^1)$, will also converge to the common limit above in a sense similar to the one that was made precise in Theorem 14.

We will illustrate the concept of modified Riemann sums in two instances, to wit, when $|J^1|$ is a function of $|J|$—linear for the example that motivated our results, and in general nonlinear—, and when $\Phi(J)$ depends on the location of J.

We begin with the former case:

Example 4 With α a fixed constant, $0 \le \alpha < 1$, let ϕ be a nonnegative function defined on $[\alpha, \alpha + |I|]$ that satisfies: (i) $\phi(t) \le t - \alpha$, for $\alpha \le t < \beta$, for some constant $\beta < \alpha + |I|$, (ii) $\phi(\alpha) = 0$, and (iii) ϕ is right-differentiable at $t = \alpha$, i.e., given $\varepsilon > 0$, there exists $\delta > 0$ so that for $\phi_+'(a)$,

$$\left| \frac{\phi(\alpha + t)}{t} - \phi_+'(\alpha) \right| \le \varepsilon, \quad \text{whenever } 0 < t < \delta. \tag{127}$$

Let the mapping Φ be defined on the subintervals J of I as follows: $\Phi(J) = J^1$, where $\chi_{\Phi(J)}$ is Riemann integrable, and $|J^1| = \phi(\alpha + |J|)$. Then, if f is Riemann–

Stieltjes integrable with respect to Ψ on I, $f\psi$ is integrable on I, and

$$u(f, \Psi) = \phi'_+(\alpha) \int_I f\psi.$$

First observe that Φ is well-defined. Indeed, for $|J| \leq \beta - \alpha$, we have $|J^1| \leq (\alpha + |J|) - \alpha = |J|$, and so we may, and do, pick $J^1 = \Phi(J) \subset J$.

Now, f is integrable with respect to Ψ on I iff $f\psi$ is integrable on I, and $\int_I f\,d\Psi = \int_I f\,\psi$. Actually the proof of the necessity, which is what is needed here, is essentially contained in the reasoning below for $\phi(t) = t$ and $\Phi(J) = J$, all J.

Let the sequence of partitions $\{\mathcal{P}_n\}$ of I satisfy simultaneously (5), hence also the equivalent of (54), for f with respect to Ψ, and (5), hence also (123), for $f\psi$, and (6) for ψ. Let N be such that $|I|/N \leq \min(\delta, \beta - \alpha)$, and consider $\mathcal{P}_n$ for $n \geq N$. If $\mathcal{P}_n$ consists of the intervals $\{I^n_k\}$, $\mathcal{P}^1_n$ is the collection $\{I^{n,1}_k\}$, with $I^{n,1}_k \subset I^n_k$, and $|I^{n,1}_k| = \phi(\alpha + |I^n_k|)$.

Observe that for $x^{n,1}_k$ in $I^{n,1}_k$, we have

$$s(f, \Psi, \mathcal{P}^1_n) = \sum_k f(x^{n,1}_k) |I^{n,1}_k|_\Psi$$

$$= \sum_k f(x^{n,1}_k) \int_{I^{n,1}_k} (\psi - \psi(x^{n,1}_k)) + \sum_k f(x^{n,1}_k)\psi(x^{n,1}_k) |I^{n,1}_k|$$

$$= A_n + B_n, \tag{128}$$

say. Now, since

$$\int_{I^{n,1}_k} |\psi - \psi(x^{n,1}_k)| \leq \mathrm{osc}\,(\psi, I^{n,1}_k) |I^{n,1}_k| \leq \mathrm{osc}\,(\psi, I^n_k) |I^n_k|,$$

with M_f a bound for f, by (6), it follows that

$$\limsup_n |A_n| \leq M_f \limsup_n \sum_k \mathrm{osc}\,(\psi, I^n_k) |I^n_k| = 0. \tag{129}$$

As for B_n, it equals

$$\sum_k f(x^{n,1}_k)\,\psi(x^{n,1}_k)\,\phi(\alpha + |I^n_k|)$$

$$= \sum_k f(x^{n,1}_k)\psi(x^{n,1}_k) \left(\frac{\phi(\alpha + |I^n_k|)}{|I^n_k|} - \phi'_+(\alpha) \right) |I^n_k|$$

$$+ \phi'_+(\alpha) \sum_k f(x^{n,1}_k)\psi(x^{n,1}_k) |I^n_k| = C_n + D_n,$$

say. Since for the intervals I_k^n in $\mathcal{P}_n$, we have $|I_k^n| \leq \delta$, by (127), C_n is bounded by

$$\left| \sum_k f(x_k^{n,1}) \, \psi(x_k^{n,1}) \left(\frac{\phi(\alpha + |I_k^n|)}{|I_k^n|} - \phi'_+(\alpha) \right) |I_k^n| \right|$$

$$\leq \sum_k |f(x_k^{n,1}) \psi(x_k^{n,1})| \left| \frac{\phi(\alpha + |I_k^n|)}{|I_k^n|} - \phi'_+(\alpha) \right| |I_k^n| \leq M_\psi \, M_f \, |I| \, \varepsilon,$$

which, since ε is arbitrary, implies that

$$\limsup_n |C_n| = 0. \tag{130}$$

Also,

$$D_n = \phi'_+(\alpha) \, S(f \, \psi, \mathcal{P}_n). \tag{131}$$

Hence, combining (128), (129), (130), and (131), it follows that

$$u(f, \Psi) = \lim_n s(f, \Psi, \mathcal{P}_n^1) = \phi'_+(\alpha) \lim_n S(f \, \psi, \mathcal{P}_n) = \phi'_+(\alpha) \int_I f \psi,$$

and we have finished.

The choice $\phi(t) = \gamma \, t$ above and the corresponding mappings Φ_γ, $0 < \gamma < 1$, which assign to an interval $J \subset I$, $\Phi_\gamma(J) = J^1 \subset I$, where χ_{J^1} is Riemann integrable on I, and J^1 is of relative length γ in J whenever $|J| < \eta$, apply to the signal retrieval result described in the introduction. As anticipated, in this case, the modified Riemann sums of f with respect Ψ on I converge to $\gamma \int_I f \, \psi$.

Along similar lines, with $\alpha \in (0, |I|]$, let ϕ be a nonnegative function defined on $[\alpha - |I|, \alpha]$ that satisfies: (i) $\phi(t) \leq \alpha - t$, for t in a left neighbourhood of α, (ii) $\phi(\alpha) = 0$, and (iii) ϕ is left-differentiable at α. Then, define $\Phi(J) = J^1$, where $|J^1| = \phi(\alpha - |J|)$. Note that $\phi(\alpha - |J|) \leq \alpha - (\alpha - |J|) = |J|$, for $|J|$ small. Then, for these values, we may, and do, pick $J^1 = \Phi(J) \subset J$ with $\chi_{\Phi(J)}$ Riemann integrable. The reader will have no difficulty in proving that in this case we have

$$u(f, \Psi) = \phi'_-(\alpha) \int_I f \, \psi.$$

The following is an instance where the mapping Φ depends on the location of the intervals J.

Example 5 Let λ be a continuous function such that $\lambda > 1$ on I, and let $0 < \gamma \leq 1$. Then, for an interval $J = [c, d] \subset I$, let $\Phi(J) = J^1$, where $J^1 = [c, c']$ is such that $\lambda(c') \int_{[c,c']} \psi = \gamma \int_I \psi$; since $c' < d$, $J^1 \subset J$. Then, if $f\lambda$ is Riemann–Stieltjes integrable with respect to Ψ on I, f is Riemann–Stieltjes integrable with respect to

Ψ on I, and

$$u(f\lambda, \Psi) = \gamma \int_I f\, d\Psi.$$

The proof of this result is left to the interested reader.

Along similar lines, there are expressions that reflect a change of variable formula for the Riemann–Stieltjes integral; general results in this direction can be found in Appendix I.

In this setting, we have two results, the first of which is:

Example 6 Let λ be a bounded, Riemann integrable function on I such that $\lambda > 1$ on I, and let Υ be an indefinite integral of $\lambda\,\psi$. Fix $0 < \gamma \le 1$, and for each interval $J = [c, d] \subset I$, pick $c' \in I$ such that $\int_{[c,c']} \lambda\,\psi = \gamma \int_J \psi$. Let Φ be defined by $\Phi(J) = J^1 = [c, c'] \subset J$. Let f be integrable with respect to Υ on I. Then, f is Riemann–Stieltjes integrable with respect to Ψ on I, and

$$u(f, \Upsilon) = \gamma \int_I f\, d\Psi.$$

The proof is left for the interested reader.

And, we consider the second result next:

Change of Variable Formula *Given a nondecreasing Lipschitz function Λ on I with Lipschitz constant 1, let $\Phi(J) = J^1 \subset I$, where, if $J = [c, d]$, $|J^1| = \Lambda(d) - \Lambda(c)$. Then, if f is Riemann–Stieltjes integrable with respect to Ψ on I, $f\psi$ is Riemann–Stieltjes integrable with respect to Λ on I, and*

$$u(f, \Psi) = \int_I f\psi\, d\Lambda.$$

Proof Note that since $(\Lambda(d) - \Lambda(c)) \le (d - c)$, it is possible to pick $J^1 \subset J$ with $\chi_{\Phi(J)}$ Riemann integrable, and we do so. Next, by the theorem in Appendix I, if f is Riemann–Stieltjes integrable with respect to Ψ on I, $f\psi$ is Riemann integrable on I, and since for $J = [c, d]$, $(\Lambda(d) - \Lambda(c)) \le (d - c)$, from (123), it follows that $f\psi$ is integrable with respect to Λ on I.

Let the sequence $\{\mathcal{P}_n\}$ of partitions of I satisfy the analogous of (5) above in this context, and hence (125), for f with respect to Ψ; the analogous of (5) above in this context, and hence (123), for $f\psi$ with respect to Λ; and (6) above for ψ. If $\mathcal{P}_n$ consists of the intervals $I_k = [x_{k,l}^n, x_{k,r}^n]$, and $\mathcal{P}_n^1$ denotes the collection $\{I_k^{n,1}\}$, we have

$$|I_k^{n,1}|_\Psi = \int_{I_k^{n,1}} \psi = \int_{I_k^{n,1}} (\psi - \psi(x_k^{n,1})) + \psi(x_k^{n,1})\big(\Lambda(x_{k,r}^n) - \Lambda(x_{k,l}^n)\big),$$

where the first summand is bounded by

$$\text{osc}\,(\psi, I_k^{n,1})\big(\Lambda(x_{k,r}^n) - \Lambda(x_{k,l}^n)\big) \le \text{osc}\,(\psi, I_k^n)\,|I_k^n|, \quad \text{all } k, n.$$

Therefore,

$$s(f, \Psi, \mathcal{P}_n^1) = \sum_k f(x_k^{n,1})\,|I_k^{n,1}|_\Psi$$

$$= \sum_k \int_{I_k^{n,1}} f(x_k^{n,1})(\psi - \psi(x_k^{n,1})) + S(f\psi, \Lambda, \mathcal{P}_n),$$

where by (6) the first sum is bounded by

$$M_f \sum_k \text{osc}\,(\psi, I_k^n)\,|I_k^n| \to 0, \quad \text{as } n \to \infty.$$

Hence,

$$u(f, \Psi) = s(f, \Psi) = \lim_n s(f, \Psi, \mathcal{P}_n^1) = \lim_n S(f\psi, \Lambda, \mathcal{P}_n) = \int_I f\psi\,d\Lambda,$$

and the proof is finished. □

We introduce now the *pattern Stieltjes* integral. As in (72), we consider the pattern sums

$$(P) \sum_k f(c_k^n)\big(\Psi(x_{k,r}^n) - \Psi(x_{k,l}^n)\big)$$

$$= \sum_{1 \le k \le n, k \in P} f(c_k^n)\big(\Psi(x_{k,r}^n) - \Psi(x_{k,l}^n)\big), \tag{132}$$

where the summation is restricted to the subset P of $\mathbb{N}$.

We then have the Principal Stieltjes Theorem:

Theorem 31 *Let $I = [a, b]$ be a bounded interval in $\mathbb{R}$, suppose that ψ and Ψ are as in (122), and let the pattern P satisfy the assumptions of the principal theorem. If the function f is Riemann–Stieltjes integrable with respect to Ψ on I, the pattern integral of f exists and*

$$(P) \int_I f\,d\Psi = \alpha \int_I f\,d\Psi. \tag{133}$$

Proof Expanding the pattern sum in (132), we have

$$\sum_{1\le k\le n, k\in P} f(c_k^n)\big(\Psi(x_{k,r}^n) - \Psi(x_{k,l}^n)\big)$$

$$= \sum_{1\le k\le n, k\in P} f(c_k^n)\Big(\int_{I_k^n}(\psi - \psi(c_k^n))\Big) + \sum_{1\le k\le n, k\in P} f(c_k^n)\psi(c_k^n)\,|I_k^n|$$

$$= A(n) + (P)\sum_k f(c_k^n)\psi(c_k^n)\,|I_k^n|, \tag{134}$$

say. Now, by (6), with M_f a bound for f, it follows that

$$\limsup_n |A(n)| \le M_f \limsup_n \sum_{k=1}^n \operatorname{osc}(\psi, I_k^n)\,|I_k^n| = 0.$$

Also, by the theorem in Appendix I, $(P)\int_I f\,\psi$ is well-defined, and taking limits in (134), by (72), and (73), we conclude that

$$(P)\int_I f\,d\Psi = (P)\int_I f\,\psi = \alpha\int_I f\,\psi. \tag{135}$$

Finally, again by the theorem in Appendix I, $\alpha\int_I f\,\psi = \alpha\int_I f\,d\Psi$, which combined with (135) yields (133) and completes the proof.   □

As for the *uniform Stieltjes integral* [91, 106], for simplicity, we will restrict ourselves to $I = [0, 1]$. We say that f is *uniformly Stieltjes integrable* on I with uniform Stieltjes integral $(SU)\int_I f\,d\Psi$, if the numerical limit

$$\lim_{c\to 0^+} \sum_{k=1}^{[1/c]} f(ck)\big(\Psi(ck) - \Psi(c(k-1))\big) = (SU)\int_I f\,d\Psi$$

exists.

We will prove two basic results, namely, that the uniform Stieltjes integral extends the Riemann–Stieltjes integral (properly) and that the uniform Stieltjes integral reduces to a uniform integral.

We begin with the former result:

Theorem 32 *Let $I = [0, 1]$, and let ψ, Ψ satisfy (122). Suppose that f is Riemann–Stieltjes integrable with respect to Ψ on I. Then, f is uniform Riemann–Stieltjes integrable on I, and*

$$(SU)\int_I f\,d\Psi = \int_I f\,d\Psi.$$

Proof For small $c > 0$, let $n = [1/c]$, and $J_k^n = [c(k-1), ck]$, $1 \leq k \leq n$; if $cn < 1$, let $J_{n+1}^n = [cn, 1]$, and if $cn = 1$, let $J_{n+1}^n = \emptyset$. Then, $Q_n = \{J_n^1, \ldots, J_{n+1}^n\}$ is a partition of $[0, 1]$, with $|J_k^n| = c$, $1 \leq k \leq n$, and, as in Theorem 21, $|J_{n+1}^n| = O(1/n)$. Then, with $c_{n+1}^n \in [cn, 1]$, we have

$$\sum_{k=1}^n f(ck)\Big(\Psi(ck) - \Psi(c(k-1))\Big) \pm f(c_{n+1}^n)\big(\Psi(1) - \Psi(cn)\big)$$

$$= S(f, \Psi, Q_n, C) + O(1/n),$$

and so,

$$(SU) \int_I f \, d\Psi = \lim_{c \to 0^+} \sum_{k=1}^{[1/c]} f(ck)\Big(\Psi(ck) - \Psi(c(k-1))\Big) = \int_I f \, d\Psi,$$

as we wanted to prove. □

In analogy with the theorem in Appendix I, we have:

Theorem 33 *Let $I = [0, 1]$, assume that ψ, Ψ are defined on I and satisfy (122), and let f be a function defined on I. Then, f is uniform Stieltjes integrable on I iff $f\psi$ is uniform integrable on I, and the uniform integrals are equal.*

Proof And by now familiar computation gives that

$$\sum_{k=1}^n f(ck)\Big(\Psi(ck) - \Psi(c(k-1))\Big) = c \sum_{k=1}^n f(ck)\psi(ck) + A(n), \tag{136}$$

where

$$A(n) = \sum_{k=1}^n f(ck)\Big(\int_{J_k^n} (\psi - \psi(ck)\Big).$$

Now,

$$|A(n)| \leq M_f \sum_{k=1}^n \operatorname{osc}(\psi, J_k^n)\,|J_k^n| \pm \operatorname{osc}(\psi, J_{n+1}^n)\,|J_k^{n+1}|$$

$$\leq (M_f + 1) \sum_{k=1}^{n+1} \operatorname{osc}(\psi, J_k^n)\,|J_k^n| + O(1/n),$$

and, therefore, since ψ is integrable, by (6) in Proposition 1, we have $\lim_n |A(n)| = 0$.

Whence, if the limit of either side of (136) exists, so does the limit in the other side and they are equal. Thus,

$$\lim_{c \to 0^+} \sum_{k=1}^{[1/c]} f(ck) \Big(\Psi(ck) - \Psi(c(k-1)) \Big) = \lim_{c \to 0^+} c \sum_{k=1}^{[1/c]} f(ck) \psi(ck),$$

and the conclusion obtains.   □

We will now discuss asymptotically Ψ-*distributed sequences mod* 1. Let Ψ : $[0, 1] \to [0, 1]$ be a monotone function with $\Psi(0) = 0$ and $\Psi(1) = 1$. A sequence $\{x_k\}$ is said to be *asymptotically Ψ-distributed mod 1* if for all $0 \le a < b \le 1$, we have

$$\lim_n \frac{1}{n} \sum_{k=1}^n \chi_{[a,b]}(x_k) = \Psi(b) - \Psi(a). \tag{137}$$

Then the following characterization holds:

Theorem 34 *A sequence $\{x_k\}$ is asymptotically Ψ-distributed mod 1 iff for all Riemann–Stieltjes integrable functions φ with respect to Ψ on $[0, 1]$,*

$$\lim_n \frac{1}{n} \sum_{k=1}^n \varphi(x_k) = \int_I \varphi \, d\Psi. \tag{138}$$

Proof We will consider necessity first. Let φ be a step function, $\varphi(x) = \sum_h c_h \chi_{J_h}$, where $\{J_h\}$ is a partition $\mathcal{P}$ of I, with $J_h = [x_{h,l}, x_{h,r}]$. Then,

$$\lim_n \frac{1}{n} \sum_{k=1}^n \sum_h c_h \chi_{J_h}(x_k) = \sum_h c_h \lim_n \frac{1}{n} \sum_{k=1}^n \chi_{J_h}(x_k)$$

$$= \sum_h c_h \Big(\Psi(x_{h,r}) - \Psi(x_{h,l}) \Big) = \int_I \varphi \, d\Psi.$$

Now, given a Riemann–Stieltjes integrable function φ with respect to Ψ and a partition $\mathcal{P} = \{J_h\}$ of I, let $m_h = \inf_{J_h} \varphi$, and $M_h = \sup_{J_h} \varphi$. Then,

$$\lim_n \frac{1}{n} \sum_{k=1}^n \sum_h m_h \chi_{J_h}(x_k) = \sum_h m_h \lim_n \frac{1}{n} \sum_{k=1}^n \chi_{J_h}(x_k) = L(\varphi, \Psi, \mathcal{P}),$$

and, similarly,

$$\lim_n \frac{1}{n} \sum_{k=1}^n \sum_h M_h \chi_{J_h}(x_k) = U(\varphi, \Psi, \mathcal{P}).$$

Moreover, since $\sum_h m_h \chi_{J_h} \leq \varphi \leq \sum_h M_h \chi_{J_h}$, throughout $[0, 1]$, it readily follows that

$$L(\varphi, \Psi, \mathcal{P}) \leq \liminf_n \frac{1}{n} \sum_{k=1}^{n} \varphi(x_k), \text{ and, } \limsup_n \frac{1}{n} \sum_{k=1}^{n} \varphi(x_k) \leq U(\varphi, \Psi, \mathcal{P}),$$

and, consequently,

$$0 \leq \limsup_n \frac{1}{n} \sum_{k=1}^{n} \varphi(x_k) - \liminf_n \frac{1}{n} \sum_{k=1}^{n} \varphi(x_k)$$

$$\leq U(\varphi, \Psi, \mathcal{P}) - L(\varphi, \Psi, \mathcal{P}),$$

which, since φ is Riemann–Stieltjes integrable with respect to Ψ, tends to 0 as $\|\mathcal{P}\| \to 0$.

Thus the limit exists. Moreover, since

$$\int_I \varphi \, d\Psi = L(\varphi, \Psi) \leq \sup_{\mathcal{P}} L(\varphi, \Psi, \mathcal{P}) \leq \lim_n \frac{1}{n} \sum_{k=1}^{n} \varphi(x_k)$$

and

$$\lim_n \frac{1}{n} \sum_{k=1}^{n} \varphi(x_k) \leq \inf_{\mathcal{P}} U(\varphi, \Psi, \mathcal{P}) = U(\varphi, \Psi) = \int_I \varphi \, d\Psi,$$

it readily follows that (137) holds, and the proof is complete.

The converse follows by setting $\varphi = \chi_{[a,b]}$ in (137). □

Finally, quadrature. For simplicity, we restrict ourselves to $I = [0, 1]$. Given Ψ that satisfies (122) and a function f defined on $(0, 1]$ that is Riemann–Stieltjes integrable with respect to Ψ on $[a, 1]$ for all $0 < a < 1$, we say that the *improper Riemann–Stieltjes integral of f with respect to Ψ* converges and has value $\int_{[0^+, 1]} f \, d\Psi$, if

$$\lim_{\varepsilon \to 0^+} \int_{[\varepsilon, 1]} f \, d\Psi = \int_{[0^+, 1]} f \, d\Psi.$$

We are interested in the extension of the sufficiency of Theorem 27 in this setting. We will begin by showing a preliminary result:

Proposition 21 *Let f be a function defined on $(0, 1]$ that is Riemann integrable on the closed subintervals $[a, b] \subset (0, 1]$, and suppose that Ψ satisfies (122). Then, f is Riemann–Stieltjes integrable with respect to Ψ on each $[a, b] \subset (0, 1]$.*

Furthermore, if the improper Riemann integral $\int_{[0^+,1]} |f|$ converges, then the improper Riemann–Stieltjes integral $\int_{[0^+,1]} f \, d\Psi$ converges.

Proof Given an interval $[a, b] \subset (0, 1]$, let $\mathcal{P} = \{I_k\}$, $I_k = [x_{k-1}, x_k]$, $1 \leq k \leq m$, be a partition of $[a, b]$. Then, since $\left|\Psi(x_k) - \Psi(x_{k-1})\right| \leq M_\psi |x_k - x_{k-1}|$, all $1 \leq k \leq m$, the Riemann–Stieltjes integrability of f with respect to Ψ on each $[a, b]$ follows readily from (124) and (6).

Thus, f is Riemann–Stieltjes integrable on each closed subinterval of I, and so the Riemann–Stieltjes integral $\int_{[\eta,1]} f \, d\Psi$ exists for every $0 < \eta \leq 1$. We claim that, given $\varepsilon > 0$, if $\eta > 0$ is sufficiently small and $0 < \eta_1 < \eta_2 < \eta$, then

$$\left| \int_{[\eta_1,\eta_2]} f \, d\Psi \right| \leq \varepsilon,$$

which will complete the proof. To see this, let $Q = \{J_1, \ldots, J_m\}$, $J_k = [x_{k-1}, x_k]$, $1 \leq k \leq m$, be a partition of $[\eta_1, \eta_2]$, and observe that with $\xi_k \in J_k$,

$$\left| \sum_{k=1}^{m} f(\xi_k)\big(\Psi(x_k) - \Psi(x_{k-1})\big) \right| \leq M_\psi \sum_{k=1}^{m} |f(\xi_k)|(x_k - x_{k-1}),$$

M_ψ being a constant. Therefore,

$$\left| \int_{[\eta_1,\eta_2]} f \, d\Psi \right| \leq M_\Psi \int_{[\eta_1,\eta_2]} |f| \leq M_\Psi \int_{[0^+,\eta]} |f|,$$

which tends to 0 as η tends to 0^+. Hence, $\int_{]0^+,1]} f \, d\Psi$ converges as long as Ψ satisfies (122), and the proof is finished. $\square$

Suppose now that $0 < \delta < 1$, that the family of partitions $\{\mathcal{P}_n\}$ satisfy (106) above, and that Ψ satisfies (122). We then say that the sequence $\{\Phi_n\}$ of functionals on the set of functions h defined on $(0, 1]$ is a *Q-sequence corresponding to* Ψ, if each Φ_n is given by

$$\Phi_n(h) = \sum_{k=1}^{m_n} w_k^n \, h(c_k^n)\big(x_k^n - x_{k-1}^n\big), \quad n = 1, 2, \ldots, \tag{139}$$

with the property that for every function f that is Riemann–Stieltjes integrable on I,

$$\lim_n \Phi_n(f) = \int_I f \, d\Psi. \tag{140}$$

Now, applying (139) and (140) to $h = \chi_I$, it readily follows that a necessary condition in this case is that

$$\lim_n \sum_{k=1}^{m_n} w_k^n \left(x_k^n - x_{k-1}^n\right) = \int_{[0,1]} \psi,$$

which we will assume. We will also assume that $|w_k^n| \le M$, for $k = 1, \ldots .m_n$, $n = 1, 2, \ldots$ We then have:

Theorem 35 *Let f be a function defined on $(0, 1]$, and suppose that Ψ satisfies (122). Then, if f is dominantly integrable, each Q-sequence $\{\Phi_n\}$ corresponding to Ψ converges, and in this case $\lim_n \Phi_n(f) = \int_{[0^+,1]} f\, d\Psi$.*

Proof First recall that by Theorem 25, the improper integral $\int_{[0^+,1]} |f|$ converges, and so, by Proposition 21, the improper Riemann–Stieltjes integral $\int_{]0^+,1]} f\, d\Psi$ also converges. Given $\varepsilon > 0$, we claim that

$$\left| \int_{[0^+,1]} f\, d\Psi - \Phi_n(f) \right| \le \varepsilon$$

provided that n is sufficiently large. We will estimate this quantity as in Theorem 27. First, since the improper Riemann–Stieltjes integral of f converges, for $\eta_1 < 1$ sufficiently small, it follows that

$$\left| \int_{[0^+,\delta\eta_1]} f\, d\Psi \right| \le \varepsilon/3. \tag{141}$$

Also for $\eta_2 < 1$, consider $\sum_{k=1}^{m_n} w_k^n \chi_{[0,\delta\eta_2]}(c_k^n) f(c_k^n)\left(x_k^n - x_{k-1}^n\right)$, which, by (110), is bounded by $\le \varepsilon/3$, for η_2 sufficiently small, independently of n. If we pick now $\eta = \min(\eta_1, \eta_2)$, (110) and (141) hold simultaneously for η, independently of n.

Finally, taking into account the above estimates,

$$\left| \int_{[0^+,1]} f\, d\Psi - \Phi_n(f) \right|$$

$$\le \left| \int_{[\delta\eta,1]} f\, d\Psi - \sum_{k=1}^{m_n} w_k^n \chi_{[\delta\eta,1]}(c_k^n) f(c_k^n)\left(x_k^n - x_{k-1}^n\right) \right| + 2\varepsilon/3,$$

where, by (140), the first summand above can be made $\le \varepsilon/3$ provided that n is large enough, and the proof is finished. $\qquad\square$

Reflecting on the learning and teaching of Mathematics, Alberto P. Calderón proposed the exercise of understanding what is a screwdriver [16]. He noted that it would be possible to write hundreds of pages describing in great detail the nature and substance of the handle, the chemical, physical, and crystalline structures of its metal parts, the detailed description of its geometric shape, the processes of preparation, and manufacture of its elements. But, having done all of this, do we really know what a screwdriver is? He suggests that we do not. To find out, it is necessary to answer the fundamental question of what a screwdriver is for. Until then, our knowledge of screwdrivers will be precarious, for understanding is only possible through the understanding of their function or purpose. And this is what we have done in this monograph: we have inspected in detail the structure of the standalone Riemann integral and illustrated its use.

Appendix I

Change of Variable Formulas for Riemann Integrals

This appendix consists of a streamlined version of the article "The change of variable formulas for Riemann integrals" that originally appeared in Real Analysis Exchange, Vol. 45, No. 1, 2020, pp. 151–172, published by Michigan State University Press. The original version can be found at https://doi.org/10.14321/realanalexch.41.1.0151.

So, we address here the change of variable, or substitution, formulas for Riemann integrals. First, we consider the general formulation by Preiss and Uher [78] of Kestelman's result pertaining the change of variable formula for the Riemann integral [24, 53]. Specifically:

Change of Variable Formula, Riemann integral *Let φ be a bounded, Riemann integrable function defined on an interval $I = [a, b]$, and let Φ be an indefinite integral of φ on I. Let f be bounded on $\Phi(I)$, the range of Φ. Then, f is Riemann integrable on $\Phi(I)$ iff $f(\Phi)\varphi$ is Riemann integrable on I, and in that case, with $\mathcal{I} = [\Phi(a), \Phi(b)]$,*

$$\int_{\mathcal{I}} f = \int_I f(\Phi)\,\varphi. \tag{142}$$

Developments in this area since Kestelman's influential paper, as well as the various strategies utilized, can be found in [6, 15, 16, 62, 71, 87, 97], and the references therein.

Since only basic properties in the theory of Riemann and Riemann–Stieltjes integration are invoked in the proof of this result, it is natural to consider a similar formula for Riemann–Stieltjes integrals. The prototype in this setting is the following:

Substitution Formula, Riemann-Stieltjes Integral *Let φ be a bounded, Riemann integrable function defined on an interval $I = [a, b]$ that does not change sign on I, and let Φ be an indefinite integral of φ on I. Let ψ be a bounded, Riemann integrable function defined on $\Phi(I)$, the range of Φ, and let Ψ be an indefinite integral of ψ on $\Phi(I)$.*

Then, if a bounded function f defined on $\Phi(I)$ is Riemann–Stieltjes integrable with respect to Ψ on $\Phi(I)$, $f(\Phi)\psi(\Phi)$ is Riemann–Stieltjes integrable with respect to Φ on I, and in that case, with $I = [\Phi(a), \Phi(b)]$,

$$\int_{\mathcal{I}} f d\Psi = \int_{I} f(\Phi)\, \psi(\Phi) d\Phi. \tag{143}$$

Note that the substitution formula holds when the Riemann–Stieltjes integral is computed with respect to an arbitrary function Ψ, and the substitution Φ is invertible. We also prove that the change of variable formula for the Riemann–Stieltjes integral holds when Ψ is monotone, or the difference of monotone functions, and Φ is not necessarily invertible. We close with the caveat that, no matter how general our results are, it is possible to obtain an instance of the change of variable formula that does not follow from them.

For the sake of making the presentation self-contained, we recall some definitions and notations. Fix a closed finite interval $I = [a, b] \subset \mathbb{R}$, and let Φ be a continuous monotone (increasing) function defined on I. For a partition $\mathcal{P}$ of I and a bounded function f on I, let $U(f, \Phi, \mathcal{P})$ and $L(f, \Phi, \mathcal{P})$ denote the upper and lower Riemann sums of f with respect to Φ on I along $\mathcal{P}$, and set $U(f, \Phi) = \inf_{\mathcal{P}} U(f, \Phi, \mathcal{P})$ and $L(f, \Phi) = \sup_{\mathcal{P}} L(f, \Phi, \mathcal{P})$.

We then say that f is Riemann–Stieltjes integrable with respect to Φ on I if $U(f, \Phi) = L(f, \Phi)$, and in this case, the common value is denoted $\int_I f\, d\Phi$, the Riemann–Stieltjes integral of f with respect to Φ on I. When $\Phi(x) = x$, one gets the usual Riemann integral on I, and Φ is omitted in the above notations. And, when it is clear from the context, integrable means Riemann–Stieltjes integrable with respect to $\Phi(x) = x$, and Riemann–Stieltjes integrable means Riemann–Stieltjes integrable with respect to a general Φ.

The following are working characterizations of Riemann–Stieltjes integrability [13, 46, 98]. A bounded function f defined on I is Riemann–Stieltjes integrable with respect to Φ on I iff, given $\varepsilon > 0$, there is a partition $\mathcal{P}$ of I, which may depend on ε, such that

$$U(f, \Phi, \mathcal{P}) - L(f, \Phi, \mathcal{P}) \le \varepsilon. \tag{144}$$

Furthermore, a sequential characterization holds, to wit, (144) is equivalent to the existence of a sequence of partitions $\{\mathcal{P}_n\}$ of I such that

$$\lim_n \big(U(f, \Phi, \mathcal{P}_n) - L(f, \Phi, \mathcal{P}_n)\big) = 0,$$

and in this case,

$$\lim_n U(f, \Phi, \mathcal{P}_n) = \lim_n L(f, \Phi, \mathcal{P}_n) = \int_I f \, d\Phi. \tag{145}$$

Integrability can also be characterized in terms of oscillations, to wit, a bounded function f is Riemann–Stieltjes integrable with respect to Φ on I iff, given $\varepsilon > 0$, there is a partition $\mathcal{P} = \{I_k\}$ of I, which may depend on ε, such that

$$\sum_k \mathrm{osc}\,(f, I_k)\left(\Phi(x_{k,r}) - \Phi(x_{k,l})\right) \le \varepsilon. \tag{146}$$

And, a sequential characterization holds, namely, (146) is equivalent to the existence of a sequence of partitions $\{\mathcal{P}_n\}$ of I consisting of the intervals $\mathcal{P}_n = \{I_k^n\}$ with $I_k^n = [x_{k,l}^n, x_{k,r}^n]$, such that

$$\lim_n \sum_k \mathrm{osc}\,(f, I_k^n)\left(\Phi(x_{k,r}^n) - \Phi(x_{k,l}^n)\right) = 0. \tag{147}$$

These characterizations do not necessarily hold if Φ fails to be monotone. Moreover, note that if (144) holds for a partition $\mathcal{P}$, it also holds for partitions $\mathcal{P}'$ finer than $\mathcal{P}$. Invoking (174) below, this observation applies to other concepts as well, including (145), (146), and (147).

Finally, since Φ is continuous and increasing on I, $\Phi(I)$ is an interval $\mathcal{I} = [\Phi(a), \Phi(b)]$ with endpoints $\Phi(a)$ and $\Phi(b)$. Note that each interval $\mathcal{J} = [y_1, y_2] \subset \mathcal{I}$ is of the form $[\Phi(x_1), \Phi(x_2)]$, where $\Phi(x_1) = y_1$, $\Phi(x_2) = y_2$, and $[x_1, x_2]$ is a subinterval of I. Moreover, partitions $\mathcal{P}$ of I induce a corresponding partition $\mathcal{Q}$ of $\mathcal{I}$, and, conversely, every partition of $\mathcal{I}$ can be expressed as $\mathcal{Q}$ for some partition $\mathcal{P}$ of I.

The Substitution Formula

We prove a result that includes the familiar substitution formula for Riemann–Stieltjes integrals [5]. The case $\Psi(x) = x$ is of some interest because for integrable f, the composition $f(\Phi)$ turns out to be Riemann–Stieltjes integrable, although $f(\Phi)$ may fail to be integrable, even if Φ is continuous [37, 53].

Proposition *Let Φ be a continuous monotone function defined on I and Ψ defined on $\mathcal{I} = \Phi(I)$. Let f be a bounded function on $\mathcal{I}$. Then, f is Riemann–Stieltjes integrable with respect to Ψ on $\mathcal{I}$ iff $f(\Phi)$ is Riemann–Stieltjes integrable with respect to $\Psi(\Phi)$ on I, and in that case, we have*

$$\int_{\mathcal{I}} f \, d\Psi = \int_I f(\Phi) \, d\Psi(\Phi). \tag{148}$$

Proof Specifically, (148) means that, if the integral on either side of the equality exists, so does the integral on the other side and they are equal. To see this, let the partition $Q = \{I_k\}$ of I correspond to the partition $\mathcal{P} = \{I_k\}$ of I such that $I_k = [\Phi(x_{k,l}), \Phi(x_{k,r})]$, where $I_k = [x_{k,l}, x_{k,r}]$. Then, since $\sup_{I_k} f = \sup_{I_k} f(\Phi)$, it readily follows that

$$U(f, \Psi, Q) = \sum_k \left(\sup_{I_k} f\right) \left(\Psi(\Phi(x_{k,r})) - \Psi(\Phi(x_{k,l}))\right)$$

$$= \sum_k \left(\sup_{I_k} f(\Phi)\right) \left(\Psi(\Phi(x_{k,r})) - \Psi(\Phi(x_{k,l}))\right) = U(f(\Phi), \Psi(\Phi), \mathcal{P}),$$

and similarly, $L(f, \Psi, Q) = L(f(\Phi), \Psi(\Phi), \mathcal{P})$. (148) follows at once from these identities.   □

Change of Variable Formula, Riemann Integral

We begin by proving a result that reduces the computation of a Riemann–Stieltjes integral to that of a Riemann integral [63]. Let φ be a bounded, Riemann integrable function defined on $I = [a, b]$, and let Φ be an indefinite integral of φ on I, i.e.,

$$\Phi(x) = \Phi(a) + \int_{[a,x]} \varphi, \quad x \in I. \tag{149}$$

We then have:

Theorem *Let Φ be as in (149) with φ positive, and let g be a bounded function on I. Then, g is Riemann–Stieltjes integrable with respect to Φ on I iff $g\,\varphi$ is integrable on I, and in that case, we have*

$$\int_I g \, d\Phi = \int_I g\varphi, \tag{150}$$

in the sense that if the integral on either side of (150) exists, so does the integral on the other side and they are equal.

Proof Assume first that g is Riemann–Stieltjes integrable, and fix $\varepsilon > 0$. Then, for a partition $\mathcal{P} = \{I_k\}$ of I with $I_k = [x_{k,l}, x_{k,r}]$, pick $\xi_k \in I_k$ such that

$$U(g\varphi, \mathcal{P}) \le \sum_k g(\xi_k)\, \varphi(\xi_k)|I_k| + \varepsilon. \tag{151}$$

There are two types of summands in (151), namely, those where $g(\xi_k) > 0$ and those where $g(\xi_k) < 0$. In the former case, note that

$$g(\xi_k)\varphi(\xi_k)|I_k| = g(\xi_k)\big(\varphi(\xi_k) - \inf_{I_k}\varphi\big)|I_k| + g(\xi_k)\big(\inf_{I_k}\varphi\big)|I_k|$$

$$\leq g(\xi_k)\,\mathrm{osc}\,(\varphi, I_k)\,|I_k| + g(\xi_k)\int_{I_k}\varphi$$

$$= g(\xi_k)\,\mathrm{osc}\,(\varphi, I_k)\,|I_k| + g(\xi_k)\big(\Phi(x_{k,r}) - \Phi(x_{k,l})\big), \qquad (152)$$

and in the latter case, since $\int_{I_k}\varphi \leq \big(\sup_{I_k}\varphi\big)|I_k|$, it follows that

$$g(\xi_k)\varphi(\xi_k)|I_k| = -g(\xi_k)\big(\sup_{I_k}\varphi - \varphi(\xi_k)\big)|I_k| + (-g(\xi_k))(-\big(\sup_{I_k}\varphi\big)|I_k|)$$

$$\leq |g(\xi_k)|\,\mathrm{osc}\,(\varphi, I_k)\,|I_k| + (-g(\xi_k))\big(-\int_{I_k}\varphi\big)$$

$$= |g(\xi_k)|\,\mathrm{osc}\,(\varphi, I_k)\,|I_k| + g(\xi_k)\big(\Phi(x_{k,r}) - \Phi(x_{k,l})\big). \qquad (153)$$

Hence, adding (152) and (153), with M_g a bound for g, we have

$$\sum_k g(\xi_k)\varphi(\xi_k)|I_k| \leq M_g \sum_k \mathrm{osc}\,(\varphi, I_k)\,|I_k| + \sum_k g(\xi_k)\big(\Phi(x_{k,r}) - \Phi(x_{k,l})\big)$$

$$\leq M_g \sum_k \mathrm{osc}\,(\varphi, I_k)\,|I_k| + U(g, \Phi, \mathcal{P}),$$

which, by (151), implies that

$$U(g\,\varphi, \mathcal{P}) \leq M_g \sum_k \mathrm{osc}\,(\varphi, I_k)\,|I_k| + U(g, \Phi, \mathcal{P}) + \varepsilon. \qquad (154)$$

Applying (154) to $-g$ gives

$$-L(g\,\varphi, \mathcal{P}) \leq M_g \sum_k \mathrm{osc}\,(\varphi, I_k)\,|I_k| - L(g, \Phi, \mathcal{P}) + \varepsilon,$$

and adding to (154), we get

$$U(g\,\varphi, \mathcal{P}) - L(g\,\varphi, \mathcal{P})$$

$$\leq 2\,M_g \sum_k \mathrm{osc}\,(\varphi, I_k)\,|I_k| + \big(U(g, \Phi, \mathcal{P}) - L(g, \Phi, \mathcal{P})\big) + 2\,\varepsilon.$$

$$(155)$$

Let $\mathcal{P}$ be a common refinement of partitions of I that satisfy (146) for φ and (144) for g with respect to Φ for the $\varepsilon > 0$ picked at the beginning of the proof; $\mathcal{P}$ then satisfies both conditions. Then, from (155), it readily follows that $U(g\,\varphi, \mathcal{P}) - L(g\,\varphi, \mathcal{P}) \le 2\, M_g\, \varepsilon + \varepsilon + 2\,\varepsilon,$, and therefore, since $\varepsilon > 0$ is arbitrary, by (144), $g\varphi$ is integrable on I.

It only remains to evaluate the integral in question. Let $\{\mathcal{P}_n\}$ be a sequence of partitions of I that satisfy simultaneously (147) for φ and (145) for g. Then, given $\varepsilon > 0$, from (154), it follows that

$$\int_I g\,\varphi = U(g\,\varphi) \le \limsup_n U(g\,\varphi, \mathcal{P}_n)$$

$$\le \limsup_n M_g \sum_k \operatorname{osc}(\varphi, I_k^n)\,|I_k^n| + \limsup_n U(g\,, \Phi\,, \mathcal{P}_n) + \varepsilon,$$

$$= \int_I g\,d\Phi + \varepsilon,$$

which, since ε is arbitrary, gives $\int_I g\,\varphi \le \int_I g\,d\Phi$. Furthermore, replacing g by $-g$, it follows that $\int_I g\,d\Phi \le \int_I g\,\varphi$, (150) holds, and the conclusion obtains.

The proof of the converse requires no new ideas, and we will be brief. Assume that $g\varphi$ is integrable on I, let $\mathcal{P} = \{I_k\}$ be a partition of I, and, given $\varepsilon > 0$, pick $\xi_k \in I_k$ such that

$$U(g, \Phi, \mathcal{P}) = \sum_k \left(\sup_{I_k} g\right)\left(\Phi(x_{k,r}) - \Phi(x_{k,l})\right) \le \sum_k g(\xi_k) \int_{I_k} \varphi + \varepsilon.$$

Proceeding as in the first part of the proof, we arrive at a relation analogous to (155), but with the Riemann sums for g and the Riemann sums for g with respect to Φ, switched, to wit,

$$U(g, \Phi, \mathcal{P}) - L(g, \Phi, \mathcal{P})$$

$$\le 2\, M_g \sum_k \operatorname{osc}(\varphi, I_k)\,|I_k| + \left(U(g\,\varphi, \mathcal{P}) - L(g\,\varphi, \mathcal{P})\right) + 2\,\varepsilon.$$

As above, we conclude that g is Riemann–Stieltjes integrable, and therefore, by the first part of the proof, $\int_I g\,d\Phi = \int_I g\,\varphi$, (150) holds, and the proof is finished.
$\square$

Proof, Change of Variable Formula, Riemann Integral We are now ready to prove the change of variable formula stated in the Introduction. It is at this juncture that we drop the assumption that φ is positive and allow it to change signs; thus, the substitution is not required to be invertible. Then $\Phi(I)$, the range of Φ, is an interval, but $\Phi(a)$, $\Phi(b)$ are not necessarily endpoints of this interval. It is important to keep in mind that the Riemann integral is oriented and that the direction in which the interval is traversed determines the sign of the integral. Also note that the assumption that f is bounded is necessary, as a simple example shows [87]. And, some care must be exercised since for $f(\Phi)\varphi$ integrable on I and φ continuous on I, it does not follow that $f(\Phi)$ is integrable on I, [53].

The proof is carried out in two parts, when φ is of constant sign, and when φ is of variable sign. In the former case, suppose first that φ is positive. Then, if f is integrable on $\mathcal{I}$, by the Proposition, $f(\Phi)$ is Riemann–Stieltjes integrable, and $\int_{\mathcal{I}} f = \int_I f(\Phi) \, d\Phi$. And, by the Theorem with $g = f(\Phi)$ there, $f(\Phi)\varphi$ is Riemann integrable on I, and $\int_I f(\Phi) \, d\Phi = \int_I f(\Phi)\varphi$. This chain of arguments shows that if f is integrable on I, $f(\Phi)\varphi$ is integrable on I, and $\int_{\mathcal{I}} f = \int_I f(\Phi)\varphi$. Moreover, since all the steps in the above argument are reversible, the converse also holds, and the change of variable formula has been established in this case.

When φ is negative, let $\psi(x) = -\varphi(x)$, and $\Psi(x) = -\Phi(x)$. Then by (142) applied to $g(x) = f(-x)$, it follows that

$$\int_{[\Psi(a),\Psi(b)]} g = \int_I g(\Psi)\,\psi, \qquad (156)$$

where the left-hand side of (156) is equal to $\int_{[-\Phi(a),-\Phi(b)]} g = -\int_{[\Phi(a),\Phi(b)]} f$, and the right-hand side equals $\int_I g(\Psi)\,\psi = \int_I g(-\Phi)\,(-\varphi) = -\int_I f(\Phi)\,\varphi$. Hence, the change of variable formula holds when φ is of constant sign, and the first part of the proof is finished.

Next, consider when φ is of variable sign. First, assume that f is integrable on $\Phi(I)$. The idea is to show that $\int_{\Phi(I)} f$ can be approximated arbitrarily close by the Riemann sums of $f(\Phi)\varphi$ on I, and, consequently, $\int_I f(\Phi)\varphi$ also exists, and the integrals are equal [6, 78]. To make this argument precise, we begin by introducing the partitions used for the approximating Riemann sums. They are based on a partition $\mathcal{P}$ of I defined as follows: given $\eta > 0$, by (146), there is a partition $\mathcal{P} = \{I_k\}$ of I, such that

$$\sum_k \operatorname{osc}(\varphi, I_k)\,|I_k| \le \eta^2 |I|. \qquad (157)$$

We first separate the indices k that appear in $\mathcal{P}$ into three classes, the (good) set G, the (bounded) set B, and the (undulating) set U, according to the following criteria. First, $k \in G$ if φ is strictly positive or negative on I_k. Next, $k \in B$, if $k \notin G$ and $|\varphi| \le \eta$ on I_k. And, finally, $k \in U$, if $k \notin G \cup B$. Note that for $k \in U$, since φ changes signs in I_k, and for at least one point ξ_k there, $|\varphi(\xi_k)| > \eta$, we have $\operatorname{osc}(\varphi, I_k) \ge \eta$.

Recall that each $I_k = [x_{k,l}, x_{k,r}]$ in $\mathcal{P}$ corresponds to the (oriented) subinterval $\mathcal{I}_k = [\Phi(x_{k,l}), \Phi(x_{k,r})]$ of $\Phi(I)$. Now, since f is integrable on $\Phi(I)$, f is integrable on $\mathcal{I}_k$, and if $k \in G$, by the first part of the proof, $f(\Phi)\varphi$ is integrable on I_k, and $\int_{\mathcal{I}_k} f = \int_{I_k} f(\Phi)\,\varphi$. Then, by (144), given $\eta > 0$, there is a partition

$\mathcal{P}^k = \{I_j^k\}$ of I_k such that

$$U(f(\Phi)\varphi, \mathcal{P}_k) - L(f(\Phi)\varphi, \mathcal{P}_k) = \sum_j \mathrm{osc}\,(f(\Phi)\varphi, I_j^k)\,|I_j^k| \le \eta\,|I_k|.$$

Moreover, since $\int_{\mathcal{I}_k} f \le U(f(\Phi)\varphi, \mathcal{P}^k)$, we also have

$$U(f(\Phi)\varphi, \mathcal{P}^k) - \int_{\mathcal{I}_k} f \le \eta\,|I_k|.$$

Hence, for $k \in G$,

$$\sum_{k \in G} \sum_j \mathrm{osc}\,(f(\Phi)\varphi, I_j^k)\,|I_j^k| \le \eta \sum_{k \in G} |I_k|, \qquad (158)$$

and

$$\sum_{k \in G} \left| \int_{\mathcal{I}^k} f - U(f(\Phi)\varphi, \mathcal{P}^k) \right| \le \eta \sum_{k \in G} |I_k|. \qquad (159)$$

Now, for $k \in B \cup U$, let $\mathcal{P}^k = \{I_k\}$ denote the partition of I_k consisting of the interval I_k. Note that, with M_φ, a bound for φ,

$$|\mathcal{I}_k| = |\Phi(x_{k,r}) - \Phi(x_{k,l})| \le \int_{[x_{k,l}, x_{k,r}]} |\varphi| \le M_\varphi\,|I_k|, \qquad (160)$$

and, with M_f, a bound for f that

$$\left| \int_{\mathcal{I}_k} f \right| \le M_f M_\varphi\,|I_k|. \qquad (161)$$

First, observe that

$$\mathrm{osc}\,(f(\Phi)\varphi, I_k)\,|I_k| = \left(\sup_{I_k} f(\Phi)\varphi - \inf_{I_k} f(\Phi)\varphi \right)|I_k| \le 2M_f M_\varphi\,|I_k|. \qquad (162)$$

Next, by (160) and (161), for $\xi_k \in I_k$,

$$\left| \int_{\mathcal{I}_k} f - f(\Phi(\xi_k))\varphi(\xi_k)\,|I_k| \right| \le 2\,M_f\,M_\varphi\,|I_k|,$$

and so, picking $\xi_k \in I_k$ appropriately, we have

$$\left| \int_{\mathcal{I}_k} f - U(f(\Phi)\varphi, \mathcal{P}^k) \right| \le 3\,M_f\,M_\varphi\,|I_k|. \qquad (163)$$

Now, if $k \in B$, $M_\varphi \leq \eta$, and therefore, from (162), we get that

$$\sum_{k \in B} \text{osc}\,(f(\Phi)\varphi, I^k)\,|I_j^k| \leq 2 M_f\, \eta \sum_{k \in B} |I_k|, \tag{164}$$

and by (163),

$$\sum_{k \in B} \left| \int_{I_k} f - U(f(\Phi)\varphi, \mathcal{P}^k) \right| \leq 3\, M_f\, \eta \sum_{k \in B} |I_k|. \tag{165}$$

Finally, since for $k \in U$, we have $\text{osc}\,(\varphi, I_k) \geq \eta$, from (157), as in Chebyshev's inequality, it follows that

$$\eta \sum_{k \in U} |I_k| \leq \sum_{k \in U} \text{osc}\,(\varphi, I_k)\,|I_k| \leq \sum_k \text{osc}\,(\varphi, I_k)\,|I_k| \leq \eta^2 |I|,$$

and, consequently,

$$\sum_{k \in U} |I_k| \leq \eta\,|I|. \tag{166}$$

Whence, by (162) and (166), the U terms are bounded by

$$\sum_{k \in U} \text{osc}\,(f(\Phi)\varphi, I^k)\,|I^k| \leq 2\, M_f\, M_\varphi \sum_{k \in U} |I_k| \leq 2\, M_f\, M_\varphi \eta\,|I|, \tag{167}$$

and by (163) and (166),

$$\sum_{k \in U} \left| \int_{I_k} f - U(f(\Phi)\varphi, \mathcal{P}^k) \right| \leq 3\, M_f M_\varphi \sum_{k \in U} |I_k| \leq 3\, M_f\, M_\varphi\, \eta\,|I|. \tag{168}$$

Consider now the partition $\mathcal{P}'$ of I that consists of the union of all the partitions $\mathcal{P}^k$, where each $\mathcal{P}^k$ is defined according as to whether $k \in G, k \in B$, or $k \in U$. Then, by (158), (164), and (167),

$$\sum_{k \in G} \sum_j \text{osc}\,(f(\Phi)\varphi, I_j^k)\,|I_j^k|$$

$$+ \sum_{k \in B} \text{osc}\,(f(\Phi)\varphi, I^k)\,|I^k| + \sum_{k \in U} \text{osc}\,(f(\Phi)\varphi, I^k)|I^k|$$

$$\leq \eta \sum_{k \in G} |I_k| + 2\, M_f\, \eta \sum_{k \in B} |I_k| + 2\, M_f M_\psi\, \eta\,|I|$$

$$\leq \left(1 + 2\, M_f + 2\, M_f M_\varphi \right) \eta\,|I|. \tag{169}$$

Given $\varepsilon > 0$, pick $\eta > 0$ so that $(1 + 2\,M_f + 2\,M_f M_\varphi)\,\eta\,|I| \le \varepsilon$, and note that the above expression is $< \varepsilon$, and since $\varepsilon > 0$ is arbitrary, (146) corresponding to $\mathcal{P}'$ implies that $f(\Phi)\varphi$ is integrable, and $L(f(\Phi)\varphi) = U(f(\Phi)\varphi) = \int_I f(\Phi)\,\varphi$.

It remains to compute the integral in question. First, note that

$$U(f(\Phi)\varphi, \mathcal{P}') = \sum_k U(f(\Phi)\varphi, \mathcal{P}^k). \tag{170}$$

Moreover, since $\Phi(b) - \Phi(a) = \sum_k \big(\Phi(x_{k,r}) - \Phi(x_{k,l})\big)$, by the linearity of the integral, taking orientation into account, it follows that $\int_I f = \sum_k \int_{I_k} f$, [81, 97]. Hence, regrouping according to the sets G, B, and U gives

$$\int_I f = \sum_{k \in G} \int_{I_k} f + \sum_{k \in B} \int_{I_k} f + \sum_{k \in U} \int_{I_k} f, \tag{171}$$

and from (170) and (171), it follows that

$$\left| \int_I f - U(f(\Phi)\varphi, \mathcal{P}') \right| \le \sum_{k \in G} \left| \int_{I_k} f - U(f(\Phi)\varphi, \mathcal{P}^k) \right|$$

$$+ \sum_{k \in B} \left| \int_{I_k} f - U(f(\Phi)\varphi, \mathcal{P}^k) \right| + \sum_{k \in U} \left| \int_{I_k} f - U(f(\Phi)\varphi, \mathcal{P}^k) \right|$$

$$= s_1 + s_2 + s_3,$$

say. Now, by (159),

$$s_1 \le \sum_{k \in G} \left| \int_{I_k} f - U(f(\Phi)\varphi, \mathcal{P}^k) \right| \le \eta \sum_{k \in G} |I_k| \le \eta\,|I|,$$

and by (165) and (168), $s_2 + s_3 \le \big(3M_f + 3M_f M_\varphi\big)\,\eta\,|I|$, which combined to give

$$\left| \int_I f - U(f(\Phi))\varphi, \mathcal{P}') \right| \le (1 + 3\,M_f + 3\,M_f M_\varphi)\,\eta\,|I|.$$

Given $\varepsilon > 0$, pick $\eta > 0$ so that $(1 + 3\,M_f + 3\,M_f M_\varphi)\,\eta\,|I| \le \varepsilon$, note that this ε also works for (169), and that

$$\left| \int_I f - U(f(\Phi)\varphi, \mathcal{P}') \right| \le \varepsilon. \tag{172}$$

Also, since $U(f(\Phi)\varphi, \mathcal{P}') - L(f(\Phi)\varphi, \mathcal{P}')$ is equal to the left-hand side of (169), from (172), it follows that

$$\left| \int_I f - L(f(\Phi)\varphi, \mathcal{P}') \right| \leq 2\varepsilon. \tag{173}$$

Furthermore, since by (172),

$$\int_I f(\Phi)\varphi = U(f(\Phi)\varphi) \leq U(f(\Phi)\varphi, \mathcal{P}') \leq \int_I f + \varepsilon,$$

and by (173),

$$\int_I f \leq L(f(\Phi)\varphi, \mathcal{P}') + 2\varepsilon \leq L(f(\Phi)\varphi) + 2\varepsilon = \int_I f(\Phi)\varphi + 2\varepsilon,$$

we conclude that

$$\left| \int_I f - \int_I f(\Phi)\varphi \right| \leq 2\varepsilon,$$

which, since ε is arbitrary, implies that $\int_I f(\Phi)\varphi = \int_I f$. Hence, (142) holds, and the proof of this implication is finished.

As for the converse, it suffices to prove that, if $f(\Phi)\varphi$ is integrable on I, f is integrable on $\Phi(I)$, for then, invoking the implication, we just proved, the integrals in question are equal. Let $\mathcal{P}$ be a partition of I that satisfies (157). Since Φ is continuous, $\Phi(I)$ is a closed interval of the form $[\Phi(x_m), \Phi(x_M)]$ with (possibly non-unique) x_m, x_M in I. If x_m and x_M are endpoints of (not necessarily the same) interval in $\mathcal{P}$, proceed. Otherwise, since for an interval $J = [x_l, x_r]$ and an interior point x of J, with $J_l = [x_l, x]$ and $J_r = [x, x_r]$, we have

$$\operatorname{osc}(\varphi, J_l)\,|J_l| + \operatorname{osc}(\varphi, J_r)\,|J_r| \leq \operatorname{osc}(\varphi, J)\,|J|; \tag{174}$$

$\mathcal{P}$ can be refined so that the endpoint that was not originally included is now an endpoint of two intervals of the new partition, without increasing the right-hand side of (178). For simplicity, also denote this new partition $\mathcal{P}$, note that it contains both x_m and x_M at least once as an endpoint of one of its intervals, and define the sets of indices G, B, and U associated to $\mathcal{P}$, as above.

Now, if $f(\Phi)\varphi$ is integrable on I, $f(\Phi)\varphi$ is integrable on I_k, and if $k \in G$, by the first part of the proof, f is integrable on I_k and $\int_{I_k} f(\Phi)\,\varphi = \int_{\mathcal{I}_k} f$. Then, by (146), given $\eta > 0$, there is a partition $Q^k = \{I_j^k\}$ of I_k, such that

$$\sum_j \operatorname{osc}(f, I_j^k)\,|I_j^k| \leq \eta\,|I_k|,$$

and therefore,

$$\sum_{k \in G} \sum_{j} \text{osc}\,(f, I_j^k)\, |I_j^k| \le \eta \sum_{k \in G} |I_k|. \tag{175}$$

As for $k \in B \cup U$, by (160), we get

$$\text{osc}\,(f, I_k)\, |I_k| = \Big(\sup_{I_k} f - \inf_{I_k} f \Big) |I_k| \le 2 M_f M_\varphi |I_k|.$$

Next, if $k \in B$, $M_\varphi \le \eta$, and therefore,

$$\sum_{k \in B} \text{osc}\,(f, I_k)\, |I_k| \le 2 M_f\, \eta \sum_{k \in B} |I_k|. \tag{176}$$

Finally, for $k \in U$, by (166), $\sum_{k \in U} |I_k| \le \eta\, |I|$, and so,

$$\sum_{k \in U} \text{osc}\,(f, I_k)|I_k| \le 2\, M_f\, M_\varphi \sum_{k \in U} |I_k| \le 2\, M_f\, M_\varphi\, \eta\, |I|. \tag{177}$$

Let Q' denote the collection of subintervals of $\Phi(I)$ defined by

$$Q' = \Big(\bigcup_{k \in G} \bigcup_{j} \{I_j^k\} \Big) \cup \Big(\bigcup_{k \in B \cup U} \{I_k\} \Big).$$

Note that the union of the intervals in Q' is $\Phi(I)$ and that, by (175), (176), and (177),

$$\sum_{k \in G} \sum_{j} \text{osc}\,(f, I_j^k)\, |I_j^k| + \sum_{k \in B} \text{osc}\,(f, I^k)\, |I^k| + \sum_{k \in U} \text{osc}\,(f, I^k)|I^k|$$

$$\le \eta \sum_{k \in G} |I_k| + 2 M_f \eta \sum_{k \in B} |I_k| + 2 M_f M_\varphi \eta\, |I|$$

$$\le \big(1 + 2 M_f + 2 M_f M_\varphi \big)\, \eta\, |I|. \tag{178}$$

Consider now the finite set $\Phi(x_m) = y_1 < y_2 < \cdots < \Phi(x_M) = y_l$, of the endpoints of the intervals in Q' arranged in an increasing fashion, without repetition. Suppose that the interval $\mathcal{J}$ in Q' contains the points $y_{k_1}, \ldots, y_{k_n}$, say, as endpoints or interior points. If they are endpoints, disregard them. Otherwise, as in (175), incorporate each, from left to right, as an endpoint of two intervals in a refined Q' without increasing the right-hand side of (145). Clearly, Q' thus refined contains a partition $Q'' = \{\mathcal{J}_k\}$ of $\Phi(I)$, which, by (178), satisfies

$$\sum_{k} \text{osc}\,(f, \mathcal{J}_k)\, |\mathcal{J}_k| \le \big(1 + 2 M_f + 2 M_f M_\varphi \big)\, \eta\, |I|.$$

Given $\varepsilon > 0$, pick $\eta > 0$ such that $(1+2M_f+2M_fM_\varphi)\,\eta\,|I| \leq \varepsilon$. Then the sum in (146) corresponding to Q'' does not exceed an arbitrary $\varepsilon > 0$, and, therefore, f is integrable on $\Phi(I)$, and the proof is finished.

Substitution Formula, Riemann–Stieltjes Integral

We begin by considering the particular case of the substitution formula when both φ and ψ are of constant sign. In this instance, we have:

Lemma *Let φ be a bounded, Riemann integrable function defined on an interval $I = [a, b]$ that does not change sign on I, and let Φ be an indefinite integral of φ on I. Let ψ be a bounded, Riemann integrable function defined on $\Phi(I)$, the range of Φ, that does not change sign, and let Ψ be an indefinite integral of ψ on $\Phi(I)$.*

Let f be bounded on $\Phi(I)$. Then, f is Riemann–Stieltjes integrable with respect to Ψ on $\Phi(I)$ iff $f(\Phi)\psi(\Phi)$ is Riemann–Stieltjes integrable with respect to Φ on I, and in that case, with $\mathcal{I} = [\Phi(a), \Phi(b)]$,

$$\int_{\mathcal{I}} f\, d\Psi = \int_I f(\Phi)\,\psi(\Phi)\,d\Phi. \tag{179}$$

Proof It suffices to prove the result when φ, ψ are positive. Indeed, if the result holds in this case, when φ is negative, it follows by replacing φ by $-\varphi$, $\psi(x)$ by $\psi(-x)$, and $f(y)$ by $f(-y)$ in (179), and when ψ is negative, by replacing ψ by $-\psi$ in (179).

Now, by assumption, we have $\Phi(x) = \Phi(a) + \int_{[a,x]} \varphi$, $x \in I$, and, $\Psi(y) = \Psi(\Phi(a)) + \int_{[\Phi(a),y]} \psi$, $y \in \Phi(I)$. A moment's thought will convince the reader that this result is akin to the theorem above, and, therefore, we shall be brief.

First, assume that $f(\Phi)\psi(\Phi)$ is Riemann–Stieltjes integrable, and fix $\varepsilon > 0$. For a partition $\mathcal{P} = \{I_k\}$ of I with $I_k = [x_{k,l}, x_{k,r}]$ and $\mathcal{I}_k = [\Phi(x_{k,r}), \Phi(x_{k,l})]$, pick $\xi_k \in I_k$ such that

$$U(f(\Phi), \Psi(\Phi), \mathcal{P}) \leq \sum_k f(\Phi(\xi_k))\left(\Psi(\Phi(x_{k,r})) - \Psi(\Phi(x_{k,l}))\right) + \varepsilon. \tag{180}$$

Then, as in the proof of the Theorem above, it follows that

$$U(f(\Phi), \Psi(\Phi), \mathcal{P}) \leq M_f \sum_k \operatorname{osc}(\psi, \mathcal{I}_k)\,|\mathcal{I}_k| + U(f(\Phi)\psi(\Phi), \Phi, \mathcal{P}) + \varepsilon, \tag{181}$$

and

$$U(f(\Phi), \Psi(\Phi), \mathcal{P}) - L(f(\Phi), \Psi(\Phi), \mathcal{P}) \leq 2M_f \sum_k \operatorname{osc}(\psi, \mathcal{I}_k)\,|\mathcal{I}_k|$$
$$+ \left(U(f(\Phi)\psi(\Phi), \Phi, \mathcal{P}) - L(f(\Phi)\psi(\Phi), \Phi, \mathcal{P})\right) + 2\varepsilon. \tag{182}$$

We pick for the partition $\mathcal{P}$ of I above a common refinement of a partition that satisfies (146) for ψ and one that satisfies (180) for $f(\Phi)\psi(\Phi)$ with respect to Φ for the $\varepsilon > 0$ picked at the beginning of the proof; $\mathcal{P}$ will then satisfy both conditions simultaneously. Then from (182), it follows that $U(f(\Phi), \Psi(\Phi), \mathcal{P}) - L(f(\Phi), \Psi(\Phi), \mathcal{P}) \leq 2M_f\varepsilon + \varepsilon + 2\varepsilon$,, and since $\varepsilon > 0$ is arbitrary, by (144), $f(\Phi)$ is Riemann–Stieltjes integrable and $U(f(\Phi), \Psi(\Phi)) = L(f(\Phi), \Psi(\Phi)) = \int_I f(\Phi)\,d\Psi(\Phi)$.

To evaluate the integral in question, as in the proof of the Theorem above with (181) playing the role of (154) there, it readily follows that $\int_I f(\Phi)d\Psi(\Phi) = \int_I f(\Phi)\psi(\Phi)\,d\Phi$. Therefore, since Φ is continuous, monotone on I, by the substitution formula (148),

$$\int_I f\,d\Psi = \int_I f(\Phi)\,d\Psi(\Phi) = \int_I f(\Phi)\psi(\Phi)\,d\Phi,$$

(179) holds, and the conclusion obtains.

To prove the converse, by (148) and the assertion we just proved, it suffices to show that, if $f(\Phi)$ is Riemann–Stieltjes integrable with respect to $\Psi(\Phi)$ on I, $f(\Phi)\psi(\Phi)$ is Riemann–Stieltjes integrable with respect to Φ on I. Let, then, $\mathcal{P} = \{I_k\}$ be a partition of I, and given $\varepsilon > 0$, pick $\xi_k \in I_k$ such that

$$U(f(\Phi)\psi(\Phi), \Phi, \mathcal{P}) \leq \sum_k f(\Phi(\xi_k))\psi(\Phi(\xi_k))\big(\Phi(x_{k,r}) - \Phi(x_{k,l})\big) + \varepsilon. \tag{183}$$

As in the proof of the Theorem above, considering the two types of summands in (183), to wit, those where $f(\Phi(\xi_k)) > 0$, and those where $f(\Phi(\xi_k)) < 0$, it follows that

$$U(f(\Phi)\psi(\Phi), \Psi(\Phi), \mathcal{P}) \leq M_f \sum_k \operatorname{osc}(\psi, I_k)\,|I_k| + U(f(\Phi), \psi(\Phi), \mathcal{P}) + \varepsilon,$$

and, consequently,

$$U(f(\Phi)\psi(\Phi), \Phi, \mathcal{P}) - L(f(\Phi)\psi(\Phi), \Phi, \mathcal{P})) \leq 2M_f \sum_k \operatorname{osc}(\psi, I_k)\,|I_k|$$
$$+ \big(U(f(\Phi), \Psi(\Phi), \mathcal{P}) - L(f(\Phi), \Psi(\Phi), \mathcal{P})\big) + 2\varepsilon,$$

and so, since $\varepsilon > 0$ is arbitrary, picking an appropriate $\mathcal{P}$, by (144), we conclude that $f(\Phi)\psi(\Phi)$ is Riemann–Stieltjes integrable, and the proof is finished. $\square$

We are now ready to prove the substitution formula stated in the Introduction. Note that in this case ψ is allowed to change signs on $\Phi(I)$.

Proof, Substitution Formula, Riemann–Stieltjes Integral The proof follows along the lines to that of the change of variable for the Riemann integral, and we shall be brief. It suffices to prove the result when φ is positive. Let f be Riemann–Stieltjes integrable with respect to Ψ on $\Phi(I)$. The idea is to show that $\int_{\Phi(I)} f \, d\Psi$ can be approximated arbitrarily close by the Riemann sums of $f(\Phi)\psi(\Phi)$ with respect to Φ on I, and, consequently, $\int_I f(\Phi)\psi(\Phi) \, d\Phi$ also exists, and the integrals are equal.

To make this argument precise, consider a partition Q of $\Phi(I)$ defined as follows: given $\eta > 0$, by (146), there is a partition $Q = \{I_k\}$ of $\Phi(I)$, such that $\sum_k \text{osc}\,(\psi, I_k)|I_k| \leq \eta^2 |I|$. We first separate the indices k that appear in Q into three classes, G, B, and U, according to the following criteria. First, $k \in G$ if ψ is strictly positive or negative on I_k. Next, $k \in B$, if $k \notin G$ and $|\psi| \leq \eta$ on I_k. And, finally, $k \in U$, if $k \notin G \cup B$. Note that for $k \in U$ we have $\text{osc}\,(\psi, I_k) \geq \eta$, since ψ changes signs in I_k and for at least one point ζ_k there, $|\psi(\zeta_k)| > \eta$.

Recall that to each subinterval $I_k = [\Phi(x_{k,l}), \Phi(x_{k,r})]$ of $\Phi(I)$ corresponds to an interval $I_k = [x_{k,l}, x_{k,r}]$, and let $\mathcal{P}$ denote the partition of I given by $\mathcal{P} = \{I_k\}$.

Now, since f is integrable with respect to Ψ on $\Phi(I)$, f is integrable with respect to Ψ on I_k, and if $k \in G$, since φ and ψ do not change sign, by the Lemma, $f(\Phi)\psi(\Phi)$ is integrable with respect to Φ on I_k, and $\int_{I_k} f \, d\Psi = \int_{I_k} f(\Phi)\,\psi(\Phi)d\Phi$. Then, by (146), given $\eta > 0$, there is a partition $\mathcal{P}^k = \{I_j^k\}$ of I_k such that

$$\sum_j \text{osc}\,(f(\Phi)\psi(\Phi), I_j^k)\,|I_j^k| \leq \eta\,|I_k|,$$

and, therefore,

$$\sum_{k \in G} \sum_j \text{osc}\,(f(\Phi)\psi(\Phi), I_j^k)\,|I_j^k| \leq \eta \sum_{k \in G} |I_k|, \qquad (184)$$

and

$$\sum_{k \in G} \left| \int_{I_k} f \, d\Psi - U(f(\Phi)\psi(\Phi), \Phi, \mathcal{P}^k) \right| \leq \eta \sum_{k \in G} |I_k|. \qquad (185)$$

Now, for $k \in B \cup U$, let $\mathcal{P}^k = \{I_k\}$ denote the partition of I_k consisting of the interval I_k. In this case, for B, we have

$$\sum_{k \in B} \text{osc}\,(f(\Phi)\psi(\Phi), I_k)\,|I_k| \leq 2\,M_f M_\varphi\, \eta \sum_{k \in B} |I_k|, \qquad (186)$$

and

$$\sum_{k \in B} \left| \int_{I_k} f \, d\Psi - U(f(\Phi)\psi(\Phi), \Phi, \mathcal{P}^k) \right| \leq 3\,M_f M_\varphi\, \eta \sum_{k \in B} |I_k|. \qquad (187)$$

and for U, with M_ψ a bound for ψ,

$$\sum_{k \in U} \mathrm{osc}\,(f(\Phi)\psi(\Phi), I_k)|I_k| \leq 2M_f M_\psi M_\varphi \sum_{k \in U} |I_k| \leq 2M_f M_\psi M_\varphi\, \eta\, |I|,$$

$$(188)$$

and

$$\sum_{k \in U} \left| \int_{I_k} f - U(f(\Phi)\psi(\Phi), \Phi, \mathcal{P}^k) \right| \leq 3\, M_f M_\psi\, M_\varphi\, \eta\, |I|. \qquad (189)$$

Consider now the partition $\mathcal{P}'$ of I that consists of the union of all the intervals in the $\mathcal{P}^k$, where each $\mathcal{P}^k$ is defined according as to whether $k \in G, k \in B$, or $k \in U$. Then, by (184), (186), and (188),

$$\sum_{k \in G} \sum_j \mathrm{osc}\,(f(\Phi)\psi(\Phi), I_j^k)\,|I_j^k|$$

$$+ \sum_{k \in B} \mathrm{osc}\,(f(\Phi)\psi(\Phi), I^k)\,|I^k| + \sum_{k \in U} \mathrm{osc}\,(f(\Phi)\psi(\Phi), I^k)|I^k|$$

$$\leq \left(1 + 2\,M_f M_\varphi + 2\,M_f M_\psi M_\varphi\right) \eta\, |I|.$$

Given $\varepsilon > 0$, pick $\eta > 0$ so that $(1+2\,M_f M_\varphi + 2\,M_f M_\psi M_\varphi)\,\eta\,|I| \leq \varepsilon$, and note that the above expression is $< \varepsilon$, and since $\varepsilon > 0$ is arbitrary and Φ is monotone, (146) corresponding to $\mathcal{P}'$ implies that $f(\Phi)\psi(\Phi)$ is Riemann–Stieltjes integrable, and $L(f(\Phi)\psi(\Phi), \Phi) = U(f(\Phi)\psi(\Phi), \Phi) = \int_I f(\Phi)\,\psi(\Phi)\,d\Phi$.

It remains to compute the integral in question. Invoking (185), (187), and (189), the computation is similar to the one carried out in the proof of the change of variable formula for the Riemann integral and is therefore omitted.

Change of Variable Formula, Riemann–Stieltjes Integral

Finally, we consider when φ is of variable sign, and in this case, the substitution is not required to be invertible. Specifically:

Change of Variable Formula, Riemann–Stieltjes Integral Let φ be a bounded, Riemann integrable function defined on an interval $I = [a, b]$, and let Φ be an indefinite integral of φ on I. Let ψ be a bounded, Riemann integrable function defined on $\Phi(I)$, the range of Φ, that does not change sign on $\Phi(I)$, and let Ψ be an indefinite integral of ψ.

Let f be a bounded function defined on $\Phi(I)$. Then, f is Riemann integrable with respect to Ψ on $\Phi(I)$ iff $f(\Phi)\psi(\Phi)$ is Riemann integrable with respect to Φ

on I, and in that case, with $\mathcal{I} = [\Phi(a), \Phi(b)]$,

$$\int_{\mathcal{I}} f \, d\Psi = \int_{I} f(\Phi) \, \psi(\Phi) d\Phi. \tag{190}$$

Proof The proof of the necessity follows along the lines of the substitution formula, and we shall be brief. It suffices to assume that ψ is positive. So, suppose that f is Riemann–Stieltjes integrable with respect to Ψ on $\Phi(I)$, and let the partition $\mathcal{P}$ of I be defined as follows: given $\eta > 0$, by (146), there is a partition $\mathcal{P} = \{I_k\}$ of I, such that

$$\sum_k \text{osc}\,(\varphi, I_k)\,|I_k| \le \eta^2 |I|. \tag{191}$$

Separate the indices k that appear in $\mathcal{P}$ into three classes, G, B, and U, according to the following criteria. First, $k \in G$ if φ is strictly positive or negative on I_k. Next, $k \in B$, if $k \notin G$ and $|\varphi| \le \eta$ on I_k. And, finally, $k \in U$, if $k \notin G \cup B$. Note that for $k \in U$, since φ changes signs in I_k and for at least one point ξ_k there, $|\varphi(\xi_k)| > \eta$, we have osc $(\varphi, I_k) \ge \eta$.

Recall that each $I_k = [x_{k,l}, x_{k,r}]$ in $\mathcal{P}$ corresponds to the (oriented) subinterval $\mathcal{I}_k = [\Phi(x_{k,l}), \Phi(x_{k,r})]$ of $\Phi(I)$. Now, since f is integrable with respect to Ψ on $\Phi(I)$, f is integrable with respect to Ψ on $\mathcal{I}_k$, and if $k \in G$, since φ and ψ do not change sign, by the lemma, $f(\Phi)\psi(\Phi)$ is integrable Φ on I_k, and by (148), $\int_{\mathcal{I}_k} f \, d\Psi = \int_{I_k} f(\Phi) \, \psi(\Phi) d\Phi$. Then, by (146), given $\eta > 0$, for each $k \in G$, there is a partition $\mathcal{P}^k = \{I_j^k\}$ of I_k, such that $\sum_j \text{osc}\,(f(\Phi)\psi(\Phi), I_j^k)\,|\mathcal{I}_j^k| \le \eta\,|I_k|$, and therefore,

$$\sum_{k \in G} \sum_j \text{osc}\,(f(\Phi)\psi(\Phi), I_j^k)\,|\mathcal{I}_j^k| \le \eta \sum_{k \in G} |I_k| \tag{192}$$

and

$$\sum_{k \in G} \left| \int_{\mathcal{I}_k} f \, d\Psi - U(f(\Phi)\psi(\Phi), \Phi, \mathcal{P}^k) \right| \le \eta \sum_{k \in G} |I_k|. \tag{193}$$

Now, for $k \in B \cup U$, let $\mathcal{P}^k = \{I_k\}$ denote the partition of I_k consisting of the interval I_k. Then, as before, it follows that

$$\sum_{k \in B} \text{osc}\,(f(\Phi)\psi(\Phi), I_k)\,|\mathcal{I}_k| \le 2\, M_f M_\psi\, \eta \sum_{k \in B} |I_k|, \tag{194}$$

and

$$\sum_{k \in B} \left| \int_{\mathcal{I}_k} f \, d\Psi - U(f(\Phi)\psi(\Phi), \Phi, \mathcal{P}^k) \right| \le 3\, M_f\, M_\psi\, \eta \sum_{k \in B} |I_k|. \tag{195}$$

Finally, for $k \in U$, as in (188) and (189), the U terms are bounded by

$$\sum_{k \in U} \operatorname{osc}\left(f(\Phi)\psi(\Phi), I_k\right)|I_k| \leq 2M_f M_\psi M_\varphi \, \eta \, |I|, \qquad (196)$$

and

$$\sum_{k \in U}\left|\int_{I_k} f - U(f(\Phi)\psi(\Phi), \Phi, \mathcal{P}^k)\right| \leq 3 \, M_f \, M_\psi \, M_\varphi \, \eta \, |I|. \qquad (197)$$

Consider now the partition $\mathcal{P}'$ of I that consists of the union of all the partitions $\mathcal{P}^k$, where each $\mathcal{P}^k$ is defined according as to whether $k \in G, k \in B$, or $k \in U$. Then, by (192), (194), and (196),

$$\sum_{k \in G}\sum_j \operatorname{osc}\left(f(\Phi)\psi(\Phi), I_j^k\right)|I_j^k| + \sum_{k \in B \cup U} \operatorname{osc}\left(f(\Phi)\psi(\Phi), I^k\right)|I^k|$$

$$\leq \left(1 + 2M_f M_\psi + 2M_f M_\psi M_\varphi\right) \eta \, |I|.$$

Given $\varepsilon > 0$, pick $\eta > 0$ so that $(1 + 2M_f M_\psi + 2M_f M_\psi M_\varphi)\,\eta\,|I| \leq \varepsilon$, and note that the above expression is $< \varepsilon$, and so, since $\varepsilon > 0$ is arbitrary, (146) corresponding to $\mathcal{P}'$ implies that $f(\Phi)\psi(\Phi)$ is Riemann–Stieltjes integrable on I and $L(f(\Phi)\psi(\Phi), \Phi) = U(f(\Phi)\psi(\Phi), \Phi) = \int_I f(\Phi)\,\psi(\Phi)\,d\Phi$.

Making use of the integral estimates (193), (195), and (197), established above, $\int_I f(\Phi)\,\psi(\Phi)\,d\Phi$ can be evaluated exactly as in the previous lemma; the computation is left to the reader.

As for the converse, it suffices to prove that, if $f(\Phi)\psi(\Phi)$ is Riemann–Stieltjes integrable with respect to Φ on I, f is Riemann–Stieltjes integrable with respect to Ψ on $\Phi(I)$, and then the integrals will be equal by the result we just proved. Let the partition $\mathcal{P}$ of I satisfy (191). Then $\mathcal{P}$ can be refined to a partition, which we call again $\mathcal{P}$ for simplicity, which contains the endpoints x_m, x_M of $\Phi(I) = [\Phi(x_m), \Phi(x_M)]$, and define the sets of indices G, B, and U, associated to $\mathcal{P}$, as above.

Now, if $f(\Phi)\psi(\Phi)$ is integrable with respect to Φ on I, $f(\Phi)\psi(\Phi)$ is integrable with respect to Φ on I_k, and if $k \in G$, since φ is of constant sign, by the lemma, f is integrable with respect to Ψ on I_k and $\int_{I_k} f(\Phi)\,\psi(\Phi)\,d\Phi = \int_{I_k} f\,d\Psi$. Then, by (146), given $\eta > 0$, there is a partition $\mathcal{Q}^k = \{I_j^k\}$ of I_k, where $I_j^k = [\Phi(x_{j,l}^k), \Phi(x_{j,r}^k)]$, such that

$$\sum_j \operatorname{osc}\left(f, I_j^k\right)\left|\Psi(\Phi(x_{j,r}^k)) - \Psi(\Phi(x_{j,l}^k))\right| \leq \eta \, |I_k|.$$

As for $k \in B \cup U$, we have

$$\sum_{k \in B} \text{osc}\,(f, \mathcal{I}_k)\,\big|\Psi(\Phi(x_{k,r})) - \Psi(\Phi(x_{k,l}))\big| \le 2M_f M_\psi\, \eta \sum_{k \in B} |I_k|,$$

and

$$\sum_{k \in U} \text{osc}\,(f, \mathcal{I}_k)\big|\Psi(\Phi(x_{k,r})) - \Psi(\Phi(x_{k,l}))\big|$$

$$\le 2\,M_f\, M_\psi\, M_\varphi \sum_{k \in U} |I_k| \le 2\,M_f\, M_\psi\, M_\varphi\, \eta\, |I|.$$

Let $\mathcal{Q}'$ denote the collection of subintervals of $\Phi(I)$ defined by

$$\mathcal{Q}' = \Big(\bigcup_{k \in G} \bigcup_{j} \{\mathcal{I}_j^k\}\Big) \cup \Big(\bigcup_{k \in B \cup U} \{\mathcal{I}_k\}\Big).$$

The proof is now concluded as that of the change of variable formula for the Riemann integral and is therefore omitted. □

Caveat

We close this appendix with a caveat: not always the most general result is the most useful. By strengthening some assumptions and weakening others, it is possible to obtain an instance of the change of variable formula that does not follow from our results [27]. We consider the Riemann–Stieltjes integral below, but this observation applies to the Riemann integral as well.

Assume that the function Φ is continuous, increasing on $I = [a, b]$, and differentiable on (a, b) with derivative $\varphi \ge 0$; then Φ is uniformly continuous on I and maps I onto $\mathcal{I} = [\Phi(a), \Phi(b)]$. Assume that Ψ is continuous, increasing on $\mathcal{I}$, and differentiable on $(\Phi(a), \Phi(b))$ with derivative $\psi \ge 0$. We will also assume that f is Riemann integrable, rather than bounded, on $\mathcal{I}$. On the other hand, we will not assume that φ, ψ are bounded. Then, if $f(\Phi)\psi(\Phi)$ is integrable with respect to Φ on I, the change of variable formula holds.

To see this, consider a partition $\mathcal{P} = \{I_k\}$, $I_k = [x_{k,l}, x_{k,r}]$, of I, and the corresponding partition $\mathcal{Q} = \{\mathcal{I}_k\}$ of $\mathcal{I}$, consisting of $\mathcal{I}_k = [y_{k,l}, y_{k,r}]$, where $y_{k,l} = \Phi(x_{k,l})$ and $y_{n,r} = \Phi(x_{k,r})$. By the mean value theorem, there exists $\zeta_k \in \mathcal{I}_k$ such that

$$\Psi(y_{k,r}) - \Psi(y_{k,l}) = \psi(\zeta_k)\,(y_{k,r} - y_{k,l}), \qquad \text{all } k,$$

and with $\xi_k \in I_k$ such that $\zeta_k = \Phi(\xi_k)$, all k, it follows that

$$\sum_k f(\zeta_k) \left(\Psi(y_{k,r}) - \Psi(y_{k,l}) \right) = \sum_k f(\Phi(\xi_k)) \, \psi(\Phi(\xi_k)) \left(\Phi(x_{k,r}) - \Phi(x_{k,l}) \right),$$

where the left-hand side is a Riemann sum of f with respect to Ψ on I, and the right-hand side a Riemann sum of $f(\Phi)\psi(\Phi)$ with respect to Φ on I. Since by the uniform continuity of Φ, it follows that $\max_k |I_k| \to 0$ implies $\max_k |\mathcal{I}_k| \to 0$, by the integrability assumptions, for appropriate partitions $\mathcal{P}$, the left-hand side above tends to $\int_{\mathcal{I}} f \, d\Psi$, and the right-hand side to $\int_I f(\Phi)\psi(\Phi) \, d\Phi$. Hence, the change of variable formula holds.

This observation applies in the following setting. On $I = \mathcal{I} = [0, 1]$, with $0 < \varepsilon, \eta < 1$, let $\Phi(x) = x^{1-\varepsilon}$, $\varphi(x) = (1 - \varepsilon) x^{-\varepsilon}$ for $x \in (0, 1]$, and $\Psi(y) = y^{1-\eta}$, $\psi(y) = (1 - \eta) y^{-\eta}$ for $y \in (0, 1]$; φ and ψ are unbounded. Then, for an integrable function f on $\mathcal{I}$, provided that $f(\Phi) \, \psi(\Phi)$ is integrable with respect to Φ on I, the change of variable formula holds. For f, we may take a continuous function of order x^β near the origin, where $\beta \geq \varepsilon/(1 - \varepsilon) + \eta$.

Along similar lines, the requirement that f be bounded on $\Phi(I)$ may be weakened in the change of variable formula for the Riemann integral and merely require that f be bounded on $[\Phi(a), \Phi(b)]$, [57]. Now, this condition is essential for the existence of the integral on the left-hand side of (142) and does not follow from the other conditions in the theorem [87].

Appendix II

Cauchy Integrability Implies Riemann Integrability

We pointed out in the Introduction that Riemann extended the notion of integral introduced by Cauchy; we come now full circle and prove that a function f that is bounded and integrable à la Cauchy, i.e., the tags of the Riemann sums of f are restricted to the left endpoints of the partition intervals, is Riemann integrable and the values of the integrals coincide. This was shown in 1915 by Gillespie [39] and has since been reproved and extended [1, 55].

Let $I = [a, b]$ and let $\mathcal{P} = \{I_k\}$, $I_k = [x_{k-1}, x_k]$, $1 \le k \le m$, be a partition of I. Recall that the left Riemann sums $S_L(f, \mathcal{P})$ of f on I are given by

$$S_L(f, \mathcal{P}) = \sum_{k=1}^{m} f(x_{k-1})(x_k - x_{k-1}).$$

We then say that f is *Cauchy integrable* on I if there is a real number C with the following property: For every $\varepsilon > 0$, there exists $\delta > 0$ such that $|S_L(f, \mathcal{P}) - C| \le \varepsilon$ for every partition $\mathcal{P}$ of I with mesh $\|\mathcal{P}\| \le \delta$. In this case, C denotes the *Cauchy integral* of f on I.

Let Π denote the admissible family of partitions of I that result by dividing I into $4n$ equal length subintervals, for all n. We then have:

Theorem *Let f be a bounded, Cauchy integrable function on an interval $I = [a, b]$. Then, given $\varepsilon > 0$, there exists $\delta > 0$ such that, given a partition $\mathcal{P}$ of I in Π with mesh $\le \delta$, there is a partition $\mathcal{Q}$ of I with mesh $\le 2\delta$ such that*

$$|S_{\Pi}(f, \mathcal{P}, T) - S_L(f, \mathcal{Q})| \le \varepsilon. \tag{198}$$

Furthermore, f is Riemann integrable on I, and $\int_I f = C$, the Cauchy integral of f.

© The Author(s), under exclusive license to Springer Nature Switzerland AG 2022
A. Torchinsky, *A Modern View of the Riemann Integral*,
Lecture Notes in Mathematics 2309, https://doi.org/10.1007/978-3-031-11799-2

Proof Given $\varepsilon > 0$, let $\eta = \varepsilon/16$. Then, since f is Cauchy integrable, there exists $\theta > 0$ such that with C the Cauchy integral of f, $\left| S_L(f, Q) - C \right| \le \eta$ for all partitions Q of I with mesh $\le \theta$. Let n_0 be an integer so that both $1/4n_0 \le \theta/2$ and, if M is a bound for f on I, $M/n_0 \le \varepsilon/2$. Then, pick $0 < \delta < 1/4n_0$. We claim that this choice of δ works.

So, let $\mathcal{P}$ be a partition of I in Π with mesh $\le \delta$, and consider the Riemann sum $S_\Pi(f, \mathcal{P}, T) = \sum_{k=1}^{4n} f(c_k)(x_k - x_{k-1})$ with tags $T = \{c_1, \ldots, c_{4n}\}$. Now, since $\mathcal{P} \in \Pi$, it follows that $(x_k - x_{k-1}) = (x_{k+1} - x_k)$ for $1 < k < 4n$, and we have the relation

$$f(c_k)(x_k - x_{k-1}) = f(c_k)(x_{k+1} - x_k)$$
$$= f(c_k)(x_{k+1} - c_{k+1}) - f(c_k)(x_k - c_k) + f(c_k)(c_{k+1} - c_k),$$

and, therefore, summing

$$\sum_{k=2}^{4n-1} \Big(f(c_k)(x_k - x_{k-1}) - f(c_k)(c_{k+1} - c_k) \Big)$$
$$= \sum_{k=2}^{4n-1} \Big(f(c_k)(x_{k+1} - c_{k+1}) - f(c_k)(x_k - c_k) \Big).$$

The idea is to prove that the sum on the right-hand side above corresponds to an error term.

Suppose first that I has been divided into 4 subintervals of equal length, and consider the left Riemann sums corresponding to the partition $\mathcal{P}_1$ with tags $\{x_0, c_1, c_2, x_2, c_3, c_4, x_4\}$, and to the partition $\mathcal{P}_2$ with tags $\{x_0, c_1, x_2, c_3, x_4\}$, and observe that

$$S_L(f, \mathcal{P}_2) - S_L(f, \mathcal{P}_1)$$
$$= \big(f(c_1) - f(c_2) \big)(x_2 - c_2) + \big(f(c_3) - f(c_4) \big)(x_4 - c_4). \qquad (199)$$

Similarly, consider the partition $\mathcal{P}_3$ with tags $\{x_0, x_1, c_2, c_3, x_3, x_4\}$ and the partition $\mathcal{P}_4$ with tags $\{x_0, x_1, c_2, x_3, x_4\}$, and observe that

$$S_L(f, \mathcal{P}_4) - S_L(f, \mathcal{P}_3) = \big(f(c_2) - f(c_3) \big)(x_3 - c_3). \qquad (200)$$

Hence, adding (199) and (200), it follows that

$$\big(S_L(f, \mathcal{P}_2) - S_L(f, \mathcal{P}_1) \big) + \big(S_L(f, \mathcal{P}_4) - S_L(f, \mathcal{P}_3) \big)$$
$$= \big(f(c_1) - f(c_2) \big)(x_2 - c_2) + \big(f(c_2) - f(c_3) \big)(x_3 - c_3)$$
$$+ \big(f(c_3) - f(c_4) \big)(x_4 - c_4),$$

which can be written as

$$f(c_1)(x_2 - c_2) + f(c_2)(x_3 - c_3) + f(c_3)(x_4 - c_4)$$
$$- \big(f(c_2)(x_2 - c_2) + f(c_3)(x_3 - c_3) + f(c_4)(x_4 - c_4)\big).$$

With this in mind, in the general case, if $\mathcal{P}_1$ has tags $\{x_0, c_1, c_2, x_2, c_3, c_4, x_4, \ldots, x_{4n}\}$, $\mathcal{P}_2$ tags $\{x_0, c_1, x_2, c_3, x_4, \ldots, x_{4n}\}$, $\mathcal{P}_3$ tags $\{x_0, x_1, c_2, c_3, x_3, c_4, c_5, \ldots, x_{4n-1}, x_{4n}\}$, and $\mathcal{P}_4$ tags $\{x_0, x_1, c_2, x_3, c_4, x_5, \ldots, x_{4n-1}, x_{4n}\}$, let

$$\mathcal{E} = \big(S_L(f, \mathcal{P}_2) - S_L(f, \mathcal{P}_1)\big) + \big(S_L(f, \mathcal{P}_4) - S_L(f, \mathcal{P}_3)\big).$$

Since $\|\mathcal{P}\| \le \delta \le \theta/2$, each of the partitions that appears in one of the left Riemann sums in the definition of $\mathcal{E}$ has mesh $\le 2\|\mathcal{P}\| \le \theta$, and so each of the left Riemann sums differs from C by less than η and, consequently,

$$|\mathcal{E}| \le 4\eta = \frac{\varepsilon}{4}. \tag{201}$$

Moreover, note that $\mathcal{E}$ can be written as

$$\mathcal{E} = \sum_{k=2}^{4n} \big(f(c_{k-1}) - f(c_k)\big)\big(x_k - c_k\big)$$

$$= \sum_{k=1}^{4n-1} f(c_k)\big(x_{k+1} - c_{k+1}\big) - \sum_{k=2}^{4n} f(c_k)\big(x_k - c_k\big). \tag{202}$$

Finally, we consider the partition Q of I consisting of the intervals

$$[x_0, c_1], [c_1, c_2], \ldots, [c_{4n-1}, c_{4n}], [c_{4n}, x_{4n}]$$

and observe that $\|Q\| \le 2\|\mathcal{P}\| \le 2\delta \le \theta$, and if

$$S_L(f, Q) = f(x_0)\big(c_1 - x_0\big)$$

$$+ \sum_{k=1}^{4n-1} f(c_k)\big(c_{k+1} - c_k\big) + f(c_{4n})\big(x_{4n} - c_{4n}\big), \tag{203}$$

it follows that $\big|S_L(f, Q) - C\big| \le \eta = \varepsilon/16$.

Now, adding (202) and (203) gives

$$\mathcal{E} + S_L(f, Q) = f(x_0)\big(c_1 - x_0\big)$$

$$+ f(c_1)(x_2 - c_1) + \sum_{k=2}^{4n-1} f(c_k)\big(x_{k+1} - x_k\big),$$

which, since $(x_{k+1} - x_k) = (x_k - x_{k-1})$, can be written as

$$\mathcal{E} + S_L(f, \mathcal{Q}) = S_\Pi(f, \mathcal{P}, T) + \mathcal{E}_1, \qquad (204)$$

where

$$\begin{aligned} \mathcal{E}_1 = f(x_0)(c_1 - x_0) + f(c_1)(x_2 - c_1) \\ - f(c_1)(x_1 - x_0) - f(c_{4n})(x_{4n} - x_{4n-1}). \end{aligned}$$

Now, if M is a bound for f on I, each of the summands on the right-hand side above is bounded by $M/4n$ and so

$$|\mathcal{E}_1| \le 4\frac{M}{4n} \le \frac{M}{n_0} \le \frac{\varepsilon}{2}. \qquad (205)$$

Finally, by (204), (201), and (205),

$$\begin{aligned} \left| S_\Pi(f, \mathcal{P}, T) - S_L(f, \mathcal{Q}) \right| &= \left| \mathcal{E} - \mathcal{E}_1 \right| \\ &\le \left| \mathcal{E} \right| + \left| \mathcal{E}_1 \right| \le \frac{\varepsilon}{4} + \frac{\varepsilon}{2} = \frac{3}{4}\varepsilon, \end{aligned}$$

(198) holds, and the first part of the proof is complete.

Moreover, with C, the Cauchy integral of f, by (198), and since $\|\mathcal{Q}\| \le \theta$, we have

$$\begin{aligned} \left| S_\Pi(f, \mathcal{P}, T) - C \right| &\le \left| S_\Pi(f, \mathcal{P}, T) - S_L(f, \mathcal{Q}) \right| + \left| S_L(f, \mathcal{Q}) - C \right| \\ &\le \frac{3}{4}\varepsilon + \eta = \frac{3}{4}\varepsilon + \frac{1}{16}\varepsilon \le \varepsilon, \end{aligned}$$

and f is Π-Riemann integrable with integral equal to C. Then, by Theorem 1, f is Riemann integrable on I with the same integral, and we have finished. $\qquad \square$

A word about the assumption of the boundedness of f. It is of course possible for unbounded functions to be Cauchy integrable. For example, if f is defined on $[0, 1)$ and is monotone in some open interval $(1 - \eta, 1)$, then the existence of the Cauchy integral is equivalent to the existence of the improper Riemann integral $\int_{[0,1-]} f(x)\,dx$. We have a function like $f(x) = (1 - x)^{-1/2}$ in mind [55].

Now, neither working with admissible families of partitions nor by only allowing left endpoints as tags enlarges the class of Riemann integrable functions. However, making both restrictions at once does. Indeed, consider the admissible class Π of partitions of $[0, 1]$ consisting of n equal length partition intervals for all n, and only left endpoints as tags. In this case, the Dirichlet function would be integrable as all its left Riemann sums are 0.

References

1. A. Abian, On a useful economy in the formation of Riemann sums. Math. Slovaca **28**(3), 247–252 (1978)
2. P. Alexandersson, N. Amini, The cone of cyclic sieving phenomena. Discrete Math. **342**(6), 1581–1601 (2019)
3. S. András, Monotonicity of certain Riemann–type sums. Teach. Math. **XV**(2), 113–120 (2012)
4. D. Andrica, M. Piticari, An extension of the Riemann–Lebesgue lemma and some applications. Acta Univ. Apulensis Math. Inform. **8**, 26–39 (2004)
5. T. Apostol, *Mathematical Analysis*, 2nd edn. (Addison-Wesley Publishing, Reading, 1974)
6. R.J. Bagby, The substitution theorem for Riemann integrals. Real Anal. Exch. **27**(1), 309–314 (2001/2002)
7. N.S. Barnett, P. Cerone, S.S. Dragomir, A sharp bound for the error in the corrected trapezoid rule and application. Tamkang J. Math. **33**(3), 253–258 (2002)
8. G. Bennett, G. Jameson, Monotonic averages of convex functions. J. Math. Anal. Appl. **252**, 410–430 (2000)
9. G.A. Bliss, A substitute for Duhamel's theorem. Ann. Math (2) **16**(1/4), 45–49 (1914–1915)
10. D. Borwein, J.M. Borwein, B. Sims, Symmetry and the monotonicity of certain Riemann sums, in *From Analysis to Visualization. JBCC 2017. Springer Proceedings in Mathematics & Statistics*, ed. by D. Bailey et al., vol. 313 (Springer, Cham, 2017)
11. A.D. Bradley, Prismatoid, prismoid, generalized prismoid. Am. Math. Monthly **86**(6), 486–490 (1979)
12. T.J. I'a Bromwich, G.H. Hardy, The definition of an infinite integral as the limit of a finite or infinite series. Q. J. Pure Appl. Math. **39**, 222–240 (1908)
13. A.M. Bruckner, J.B. Bruckner, B.S. Thomson, *Real Analysis*, 2nd edn. (ClassicRealAnalysis.com, 2008)
14. N.G. de Bruijn, K.A. Post, A remark on uniformly distributed sequences and Riemann integrability. Indag. Math. **30**, 149–150 (1968)
15. J.S. Byrnes, A. Giroux, O. Shisha, Riemann sums and improper integrals of step functions related to the prime number theorem. J. Approx. Theory **40**, 180–192 (1984)
16. A.P. Calderón, Reflexiones sobre el aprendizaje y la enseñanza de las matemáticas. Revista de didáctica de matemáticas **33**, 3–12 (1998)
17. R.E. Carr, J.D. Hill, Pattern integration. Proc. Am. Math. Soc. **2**, 242–245 (1951)
18. D.M. Cates, *Cauchy's Calcul Infinitésimal: An Annotated English translation* (Springer, Berlin, 2019)

© The Author(s), under exclusive license to Springer Nature Switzerland AG 2022
A. Torchinsky, *A Modern View of the Riemann Integral*,
Lecture Notes in Mathematics 2309, https://doi.org/10.1007/978-3-031-11799-2

19. C.-H. Ching, C.K. Chui, Asymptotic similarities of Fourier and Riemann coefficients. J. Approx. Theory **10**, 295–300 (1974)
20. C. Chui, Concerning rates of convergence of Riemann sums. J. Approx. Theory **4**, 279–287 (1971)
21. O. Costin, N. Falkner, J.D. McNeal, Some generalizations of the Riemann-Lebesgue lemma. Am. Math. Monthly **123**(4), 387–391 (2016)
22. D. Cruz Uribe, C.J. Neugebauer, Sharp error bounds for the trapezoidal rule and Simpson's rule. J. Ineq. Pure Appl. Math. **3**(4, Article 49) (2002)
23. D. Cruz Uribe, C.J. Neugebauer, An elementary proof of error estimates for the trapezoidal rule. Math. Mag. **76**, 303—306 (2003)
24. R.O. Davies, An elementary proof of the theorem on change of variable in Riemann integration. Math. Gaz. **45**, 23–25 (1961)
25. P.J. Davis, P. Rabinowitz, *Methods of Numerical Integration*, 2nd edn. (Academic Press, New York, 2014)
26. R.E. Deakin, M.N. Hunter, The prismoidal correction revisited. Presented at 22nd Victorian Regional Surveying Conference, Beechworth, Victoria, 20–21 February, 2009
27. O.R.B. de Oliveira, Change of variable for the Riemann integral on the real line. Preprint (2019)
28. D. Dickinson, Approximative Riemann–sums for improper integrals. Q. J. Math. **12**(1), 176–183 (1941)
29. T.S. dos Reis, Proper and improper Riemann integral in a single definition. Proc. Series Braz. Soc. Appl. Comput. Math. **5**(1), 1–7 (2017)
30. S.S. Dragomir, R.P. Agarwal, P. Cerone, On Simpson's inequality and applications. J. Inequal. Appl. **5**(6), 533—579 (2000)
31. S.S. Dragomir, P. Cerone, A. Sofo, Some remarks on the trapezoid rule in numerical integration. Indian J. Pure Appl. Math. **31** , 475–494 (2000)
32. H.J. Ettlinger, A simple form of Duhamel's theorem and some new applications. Am. Math. Monthly **29**(7), 239–250 (1922)
33. T.H. Fay, P.H. Kloppers, The Gibbs' phenomenon. Int. J. Math. Educ. Sci. Technol. **32**(1), 873–899 (2001)
34. E.C. Fazekas Jr., P.R. Mercer, Elementary proofs of error estimates for the midpoint and Simpson's rules. Math. Mag. **82**(5), 365–370 (2009)
35. R.L. Foote, H. Nie, How to approximate the volume of a lake. College Math. J. **47**(3), 162–170 (2016)
36. M. Frontini, E. Sormani, Some variant of Newton's method with third–order convergence. Appl. Math. Comput. **140**, 419—426 (2003)
37. B.R. Gelbaum, J.M.H. Olmsted, *Counterexamples in Analysis* (Holden-Day, San Francisco, 1965)
38. A. Ghizzetti, A. Ossicini, *Quadrature Formulae*. International Series of Numerical Mathematics, vol. 13 (Birkhauser Verlag, Basel-Stuttgart, 1970)
39. R.C. Gillespie, The Cauchy definition of a definite integral. Ann. Math (2) **17**(2), 61–63 (1915)
40. R.J. Gillins, The volume of a truncated pyramid in ancient Egyptian papyri. Math. Teach. **57**(8), 552–555 (1964)
41. E.A. González–Velasco, The Lebesgue integral as a Riemann integral. Int. J. Math. Math. Sci. **10**(4), 693–706 (1987)
42. S. Haber, O. Shisha, Improper integrals, simple integrals, and numerical quadrature. J. Approx. Theory **11**, 1–15 (1974)
43. R.W. Hamming, Mathematics on a distant planet. Am. Math. Monthly **105**(7), 640–650 (1998)
44. H. Hankel, Untersuchungen über die unendlich oft oszillierenden und unstetigen funktionen. Reprinted Math. Ann. **20**(1), 63–112 (1882)
45. J.D. Hill, Summability methods defined by Riemann sums. Can. J. Math. **5** , 289–296 (1953)
46. J.K. Hunter, *Introduction to Analysis*. Undergraduate Lecture Notes (U C Davis, 2010)

47. A. Imhausen, *Mathematics in Ancient Egypt: A Contextual History* (Princeton University Press, 2016)
48. A.E. Ingham, Improper integrals as limits of sums. J. Lond. Math. Soc. **24**, 44–50 (1949)
49. D. Jackson, *The Theory of Approximation*, vol. 11 (American Mathematical Society Colloquium Publications, New York). Published by AMS, 1930
50. S.G. Johnson, Numerical integration and the redemption of the trapezoidal rule. Preprint
51. C.S. Kahane, Generalizations of the Riemann-Lebesgue and Cantor-Lebesgue lemmas. Czech. Math. J., **30**(105), 108–117 (1980)
52. C.X. Kang, E. Yi, Disk versus frustum. Texas College Math. J. **4**(2), 13–20 (2007)
53. H. Kestelman, Change of variable in Riemann integration. Math. Gaz. **45**, 17–23 (1961)
54. K. Kirkpatrick, E. Lenzmann, G. Staffilani, On the continuum limit for discrete NLS with long–range lattice interactions. Commun. Math. Phys. **317**, 563–591 (2013)
55. E. Kristensen, E.T. Poulsen, E. Reich, A characterization of Riemann–integrability. Am. Math. Monthly **69**(6), 498–505 (1962)
56. L. Kuipers, H. Niederreiter, *Uniform Distribution of Sequences* (Wiley-Interscience John Wiley and Sons, New York-London-Sydney, 1974)
57. A. Kuleshov, A remark on the change of variable theorem for the Riemann integral. Mathematics **9**(16), 1899 (2021)
58. I. Kyrezi, Monotonicity properties of Darboux sums. Real Anal. Exch. **35**(1), 43–64 (2009/2010)
59. J.T. Lewis, C.F. Osgood, O. Shisha, Infinite Riemann sums, the simple integral, and the dominated integral, in *General Inequalities 1/Allgemeine Ungleichungen 1*. International Series of Numerical Mathematics/Internationale Schriftenreihe zur Numerischen Mathematik/Série Internationale d'Analyse Numérique, ed. by E.F. Beckenbach, vol 41 (Birkhäuser, Basel, 1978), pp. 233–242
60. J. Liu, Approximative theorem of incomplete Riemann–Stieltjes sum of stochastic integral (2018). arXiv:1803.05182v2 [math.CA]
61. J. Liu, Y. Liu, Deleting items and disturbing mesh theorems for Riemann definite integral and their applications. arXiv:1702.04464v1 [math.CA] (2017)
62. R. López Pouso, Riemann integration via primitives for a new proof to the change of variable theorem (2011). arXiv:1105.5938v1 [math.CA]
63. R. López Pouso, Existence and computation of Riemann–Stieltjes integrals through Riemann integrals (2011). arXiv:1107.1996v1 [math.CA]
64. B. Lu, D.H. Torchinsky, Fourier domain rotational anisotropy second harmonic generation. Opt. Exp. **26**(25), 33192–33204 (2018)
65. J. Lu, Is the composite function integrable? Am. Math. Monthly **106**(8), 763–766 (1999)
66. T. Malvic, K. Novak Zelenika, Why we use Simpson and trapezoidal rule for hydrocarbon reservoir volume calculation? in *Congress Book "Geomathematics–From Theory to Practice*, ed. by M. Cvetković, K. Novak Zelenika, J. Geiger (Croatian Geological Society, 2014, Opatija), pp. 37–44
67. D. Ş. Marinescu, M. Monea, An extension of a result about the order of convergence. Bull. Math. Anal. Appl. **3**(3), 25–34 (2011)
68. E.J. McShane, Partial orderings and Moore–Smith limits. Am. Math. Monthly **59**, 1–11 (1952)
69. P.R. Mercer, *More Calculus of a Single Variable*. Undergraduate Texts in Mathematics (Springer, Berlin, 2014)
70. B.E. Meserve, R.E. Pingry, Some notes on the prismoidal formula. Math. Teach. **45**(4), 257–263 (1952)
71. J. Navrátil, A note on the theorem on change of variable in a Riemann integral (Czech). Časopis Pěst. Mat. **106**(1), 79—83 (1981)
72. C.F. Ogood, O. Shisha, The dominated integral. J. Approx. Theory **17**, 150–165 (1976)

73. C.F. Osgood, O. Shisha, Numerical quadrature of improper integrals and the dominated integral. J. Approx. Theory **20**, 139–152 (1977)
74. W.F. Osgood, The integral as the limit of a sum, and a theorem of Duhamel's. Ann. Math (2) **4**(4), 161–178 (1903)
75. L. Owens, Exploring the rate of convergence of approximations to the Riemann integral. Preprint
76. J. Pecaric, S. Varosanec, Simpson's formula for functions whose derivatives belong to L_p spaces. App. Math. Lett. **14**(2), 131–135 (2001)
77. A. Pinkus, Weierstrass and approximation Theory. J. Approx. Theory **107**(1), 1–66 (2000)
78. D. Preiss, J. Uher, A remark on the substitution for the Riemann integral. (Czech) Časopis Pěst. Mat. **95**, 345—347 (1970)
79. P. Renz, The path to hell.... College Math. J. **16**(1), 9–11 (1985)
80. J. Rey Pastor, *Elementos de la Teoría de Funciones*, 3rd. edn. (Ibero-Americana, Madrid, 1953)
81. H.E. Robbins, A note on the Riemann integral. Am. Math. Monthly **50**(10), 617–618 (1943)
82. D. Ruch, The definite integrals of Cauchy and Riemann (2017). Analysis. 11. https:// digitalcommons.ursinus.edu/triumphs_analysis/11
83. J.J. Ruch, M. Weber, Quelques résultats dans la téorie des sommes riemanniennes. Exposition. Math. **15**(3), 279–288 (1997)
84. S. Salvati, A. Volčič, A quantitative version of a de Bruijn–Post theorem. Math. Nachr. **229**, 161–173 (2001)
85. R. Sandberg, On the compatibility of the uniform integral. Notices Am. Math. Soc. 67, 265 (1963)
86. R. Sandberg, *On the Compatibility of Integrals* (Dissertation, University of Arizona, 1964)
87. D.N. Sarkhel, R. Výborný, A change of variables theorem for the Riemann integral. Real Anal. Exch. **22**(1), 390—395 (1996/1997)
88. S. Selvaraj, A note on Riemann sums and improper integrals related to the prime number theorem. J. Approx. Theory **66**, 106–108 (1991)
89. J.A. Shohat, Definite integrals and Riemann sums. Am. Math. Monthly **46**, 538–545 (1939)
90. A. Sklar, On the definition of the Riemann integral. Am. Math. Monthly **67**(9), 897–900 (1960)
91. A. Sklar, Uniform Stieltjes integral. Notices Am. Math. Soc., **11**, 611–685 (1964)
92. H.J.S. Smith, On the integration of discontinuous functions. Proc. Lond. Math. Soc. **6**, 140—153 (1875)
93. L.A. Talman, Simpson's rule is exact for quintics. Am. Math. Monthly **113**(2), 144–155 (2006)
94. H. Tandra, A new proof of the change of variable theorem for the Riemann integral. Am. Math. Monthly **122**(8), 795–799 (2015). https://doi.org/10.4169/amer.math.monthly.122.8.795
95. H. Tandra, Corrigendum to "A new proof of the change of variable theorem for the Riemann integral." Am. Math. Monthly **123**(10), 1049 (2016)
96. B.S. Thomson, Characterizations of an indefinite Riemann integral. Real Anal. Exch. **35**(2), 487–492 (2009)
97. B.S. Thomson, On Riemann sums. Real Anal. Exch. **37**(1), 221–242 (2011/2012)
98. A. Torchinsky, *Real Variables* (Addison-Wesley Publishing, Redwood City, 1988)
99. A. Torchinsky, *Problems in Real and Functional Analysis*. Graduate Studies in Mathematics, vol. 166 (American Mathematical Society, Providence, 2015)
100. A. Torchinsky, Modified Riemann sums of Riemann–Stieltjes integrable functions (2019). arXiv: 1905.00881v1[math.CA]
101. A. Torchinsky, The change of variable formulas for Riemann integrals. Real Anal. Exch. **45**(1), 151–172 (2020)
102. W.F. Trench, A Riemann integral proof of a generalized Riemann lemma. Preprint
103. N. Ujevic, Sharp inequalities of Simpson type and Ostrowski type. Comput. Math. Appl. 48, 145—151 (2004)

104. V. Volterra, Alcune osservasioni sulle funzioni punteggiate discontinue, in *Opere Matematiche 1* (Rome 1954), 7–15 (Originally in Giornale di Matematiche, vol. 19 (1881), 76–86
105. J.L. Wals, W.E. Sewell, Note on degree of approximation to an integral by Riemann sums. Am. Math. Monthly **44**(3), 155–160 (1937)
106. B.J. Walsh, *The Uniform and uniform Stieltjes integrals*. Dissertations, Theses, and Master's Projects. Paper 1539624580, 1965
107. S. Weerakoon, T.G.I. Fernando, A variant of Newton's method with accelerated third-order convergence. Appl. Math. Lett. **13**(8), 87—93 (2000)
108. A. Wintner, The sum formula of Euler–Maclaurin and the inversions of Fourier and Möebius. Am. J. of Math. **69**(4), 685–708 (1947)

Index

© The Author(s), under exclusive license to Springer Nature Switzerland AG 2022
A. Torchinsky, *A Modern View of the Riemann Integral*,
Lecture Notes in Mathematics 2309, https://doi.org/10.1007/978-3-031-11799-2

Printed in the United States
by Baker & Taylor Publisher Services